Sir Horace Lamb

A treatise on the mathematical theory of the motion of fluids

Sir Horace Lamb

A treatise on the mathematical theory of the motion of fluids

ISBN/EAN: 9783742892294

Manufactured in Europe, USA, Canada, Australia, Japa

Cover: Foto ©berggeist007 / pixelio.de

Manufactured and distributed by brebook publishing software
(www.brebook.com)

Sir Horace Lamb

A treatise on the mathematical theory of the motion of fluids

A TREATISE

ON THE

MATHEMATICAL THEORY

OF

THE MOTION OF FLUIDS,

BY

HORACE LAMB, M.A.

FORMERLY FELLOW AND ASSISTANT TUTOR OF TRINITY COLLEGE, CAMBRIDGE;
PROFESSOR OF MATHEMATICS IN THE UNIVERSITY OF ADELAIDE.

Cambridge:
AT THE UNIVERSITY PRESS.

London: CAMBRIDGE WAREHOUSE, 17, PATERNOSTER ROW.
Cambridge: DEIGHTON, BELL, AND CO.
Leipzig: F. A. BROCKHAUS.

1879

PREFACE.

THE following attempt to set forth in a systematic and connected form the present state of the theory of the Motion of Fluids, had its origin in a course of lectures delivered in Trinity College, Cambridge, in 1874, when the need for a treatise on the subject was strongly impressed on my mind. Various circumstances have retarded the completion of the work in a form fit for the press; but as the delay has enabled me to incorporate the results of several important recent investigations, and altogether to render the volume less inadequate to its purpose than it would otherwise have been, this is hardly matter for regret.

I have endeavoured, throughout the book, to attribute to their proper authors the various steps in the development of the subject. The list of Memoirs and Treatises at the end of the book has no pretensions to completeness, and as it is to a great extent based on

MS. notes which I have no present means of verifying, some of the references may possibly be inexact. I trust however that the list may, in spite of these drawbacks, be of service to the student who wishes to consult the original authorities.

I am under great obligations to my friends Mr H. M. Taylor and Mr W. D. Niven of Trinity College for their kindness in correcting the proof-sheets and in generally supervising the passage of the work through the press.

HORACE LAMB.

ADELAIDE,
May 16, 1879.

CONTENTS.

CHAPTER I.

THE EQUATIONS OF MOTION.

ART.		PAGE
1—3.	Fundamental assumptions	1
4.	Two forms of the equations	3
5—11.	The Eulerian form of the equations. Dynamical equations, equation of continuity, surface conditions	3
12—14.	Second method. Flow of matter, momentum, energy . . .	8
15.	Impulsive generation of motion	11
16, 17.	The Lagrangian form of the equations. Dynamical equations, equation of continuity	12
18—21.	Weber's Transformation.	14
	Comparison of the Eulerian and Lagrangian methods.	

CHAPTER II.

INTEGRATION OF THE EQUATIONS IN SPECIAL CASES.

22—27.	Velocity-Potential. Lagrange's Theorem. Equi-potential surfaces. Physical interpretation of velocity potential . .	18
28—30.	Steady Motion	22
31—37.	Examples: Efflux of liquids and gases. The Vena Contracta. Rotating fluid	25

CHAPTER III.

IRROTATIONAL MOTION.

38—40.	Analysis of the motion of a fluid element	32
	'Circulation.'	
41, 42.	Irrotational motion in simply-connected spaces	37
43—52.	Particular case of a liquid. Properties of the velocity potential .	38
53—59.	Multiply-connected spaces	47
	Irrotational motion in such spaces.	

ART.		PAGE.
60—63.	Particular case of a liquid	. 54
64, 65.	Green's Theorem	. 57
	Kinetic energy of a fluid.	
66, 67.	Thomson's extension of Green's Theorem	63

CHAPTER IV.

MOTION OF A LIQUID IN TWO DIMENSIONS.

68, 69.	The Stream-Function	. 66
70—72.	Irrotational motion. Kinetic energy	. 68
73—79.	Relations between the velocity potential and the stream-function .	71
	Digression on the theory of functions of a complex variable.	
	Logarithmic function.	
80—84.	Examples. Special cases of motion . . . . , , .	80
85—87.	General formulæ for the velocity potential and the stream-function	84
	Motion of a circular cylinder in a liquid.	
88.	Motion of translation of a solid through a liquid	87
	Examples.	
89.	Motion of rotation of a solid in a liquid .	91
	Examples. ·	
90—93.	Inverse complex functions .	94
	Kirchhoff's method.	
	Examples.	
94—98.	Discontinuous Motions	98
	Helmholtz's and Kirchhoff's solutions.	
	The Vena Contracta.	
	Impact of a stream on a lamina.	

CHAPTER V.

MOTION OF SOLIDS THROUGH A LIQUID.

99—104.	Kinematical investigations	. 110
	Examples.	
105.	Dynamical investigations	. 118
	Direct method. Application to a sphere.	
106—108.	Indirect method	. 120
	'Impulse.'	
109.	Equations of motion	. 123
110—112.	Kinetic energy. Momenta	. 124
113, 114.	Cases of steady motion	. 126
115.	Motion of a solid struck by an impulsive couple . . .	. 129
116.	Expressions for the kinetic energy in certain cases. . .	. 130
117.	Motion of a solid of revolution in a particular case. . .	. 133
118.	Stability of a body moving parallel to an axis of symmetry .	. 134
119.	Constraint necessary to produce any specified motion . .	. 135

ART.   PAGE
120, 121.   Case of a perforated solid.   Cyclic motion.   .   .   136
122, 123.   Motion of a ring   140
124.   Case of two or more solids   141
125.   Case of two spheres   142

CHAPTER VI.

VORTEX MOTION.

126, 127.   Definitions.   'Strength' of a vortex   146
128—131.   Determination of motion in terms of expansion and rotation   .   149
            Interpretation.   Electromagnetic analogy.
132.   Vortex-Sheet   154
133.   Velocity-Potential due to a vortex   156
134.   Dynamical theorems   157
135—137.   Kinetic Energy   159
138, 139.   Rectilinear vortices   162
            Examples.
140—144.   Circular vortices   167
145.   Conditions for steady motion   172

CHAPTER VII.

WAVES IN LIQUIDS.

146—152.   Waves of small vertical displacement   .   .   .   173
            Waves in canals.
153—155.   Small disturbing forces   182
            Canal theory of the tides.
156—160.   Waves in deep water   185
161—163.   Wave-propagation in two dimensions.   Examples   .   194
164.   Free oscillations in an ocean of uniform depth.   .   .   197

CHAPTER VIII.

WAVES IN AIR.

165, 166.   Plane waves   201
167, 168.   Spherical waves   203
169, 170.   General equations of sound waves.   Integration .   .   206
171, 172.   Solutions of the exact equations of motion of plane waves   .   208
            Methods of Earnshaw and Riemann.
            Change of type.
173.   Condition for permanency of type   213

CHAPTER IX.

VISCOSITY.

ART. PAGE

174, 175. Specification of the state of stress at any point of a viscous fluid . 215
 Equations of motion.
176, 177. Formulæ of transformation 216
 Relations between the stresses across different planes.
178. Stresses in terms of rates of strain 218
 Coefficient of viscosity.
179. Dissipation-function 220
180, 181. Equations of motion 221
 Boundary-conditions.
182—185. Examples: Flux through a straight tube . 223
 Effect of viscosity on motion of sound.
 Uniform motion of a sphere.

NOTES.

A. Establishment of the fundamental equations on the molecular hypothesis 231

B. Multiply-connected regions. Definitions self-consistent. Barriers . . 237

C. Formulæ for component momenta of a solid immersed in a liquid . . 239

D. Vortex-motion . . . 241

E. On the resistance of fluids . 244

LIST OF MEMOIRS AND TREATISES . . 249
EXERCISES . . . 252

ON FLUID MOTION.

CHAPTER I.

1. THE following investigations proceed on the assumption that the fluids with which we deal may be treated as practically continuous and homogeneous in structure; *i.e.* we assume that the properties of the smallest portions into which we can conceive them to be divided are the same as those of the substance in bulk. It is shewn in note (A), at the end of the book, that the fundamental equations arrived at on this supposition, with proper modifications of the meanings of the symbols, still hold when we take account of the heterogeneous or molecular structure which is most probably possessed by all ordinary matter.

2. The fundamental property of a fluid is that it cannot be in equilibrium in a state of stress such that the mutual action between two adjacent parts is oblique to the common surface. This property is the basis of Hydrostatics, and is verified by the complete agreement of the deductions of that science with experiment. Very slight observation is enough, however, to convince us that oblique stresses may exist in fluids *in motion*. Let us suppose for instance that a vessel in the form of a circular cylinder, containing water (or other liquid), is made to rotate about its axis, which is vertical. If the motion of the vessel be uniform, the fluid is soon found to be rotating with the vessel as one solid body. If the vessel be now brought to rest, the motion of the fluid continues for some time, but gradually subsides, and

L.   1

at length ceases altogether; and it is found that during this process the portions of fluid which are further from the axis lag behind those which are nearer, and have their motion more rapidly checked. These phenomena point to the existence of mutual actions between contiguous elements which are partly tangential to the common surface. For if the mutual action were everywhere wholly normal, it is obvious that the moment of momentum, about the axis of the vessel, of any portion of fluid bounded by a surface of revolution about this axis, would be constant. We infer, moreover, that these tangential stresses are not called into play so long as the fluid moves as a solid body, but only whilst a change of shape of some portion of the mass is going on, and that their tendency is to oppose this change of shape.

3.   It is usual, however, in the first instance, to neglect the tangential stresses altogether. Their effect is in many practical cases small, but, independently of this, it is convenient to divide the not inconsiderable difficulties of our subject by investigating first the effects of purely normal stress. The further consideration of the laws of tangential stress is accordingly deferred till Chapter IX.

If the stress exerted across any small plane area situated at a point P of the fluid be wholly normal, its intensity (per unit area) is the same for all aspects of the plane. The following proof of

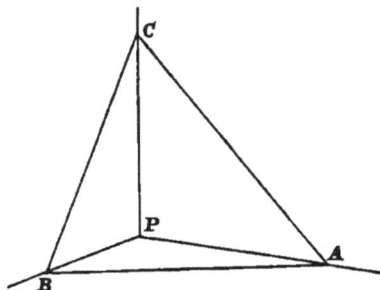

this theorem is given here for purposes of reference. Through P draw three straight lines PA, PB, PC mutually at right angles, and let a plane whose direction-cosines relatively to these lines are l, m, n, passing infinitely close to P, meet them in A, B, C.

Let p, p_1, p_2, p_3 denote the intensities of the stresses * across the faces ABC, PBC, PCA, PAB, respectively, of the tetrahedron $PABC$. If Δ be the area of the first-mentioned face, the areas of the others in order are $l\Delta$, $m\Delta$, $n\Delta$. Hence if we form the equation of motion of the tetrahedron parallel to PA we have

$$p_1 \cdot l\Delta = pl \cdot \Delta,$$

where we have omitted the terms which express the rate of change of momentum, and the component of the external impressed forces, because they are ultimately proportional to the mass of the tetrahedron, and therefore of the third order of small quantities, whilst the terms retained in the equation of motion are of the second. We have then, ultimately, $p = p_1$, and similarly $p = p_2 = p_3$, which proves the theorem.

4. The equations of motion of a fluid have been obtained in two different forms, corresponding to the two ways in which the problem of determining the motion of a fluid mass, acted on by given forces and subject to given conditions, may be viewed. We may either regard as the object of our investigations a knowledge of the velocity, the pressure, and the density, at all points of space occupied by the fluid, for all instants; or we may seek to determine the history of each individual particle. The equations obtained on these two plans are conveniently designated, as by German mathematicians, the 'Eulerian' and the 'Lagrangian' forms of the hydrokinetic equations, although both forms are in reality due to Euler †.

The Eulerian Forms of the Equations.

5. Let u, v, w be the components, parallel to the co-ordinate axes, of the velocity at the point (x, y, z) at the time t. These quantities are then functions of the independent variables x, y, z, t. For any particular value of t they express the motion at that

---

* Reckoned positive when pressures, negative when tensions. Ordinary fluids are, however, incapable of supporting more than an exceedingly slight degree of tension, so that p is nearly always positive.

† Principes généraux du mouvement des fluides. *Hist. de l'Acad. de Berlin*, 1755.

De principiis motus fluidorum. *Novi Comm. Acad. Petrop.* t. 14, p. 1, 1759.

Lagrange starts in the *Mécanique Analytique* with the second form of the equations, but transforms them at once to the 'Eulerian' form.

instant at all points of space occupied by the fluid; whilst for particular values of x, y, z they give the history of what goes on at a particular place.

Now let F be any function of x, y, z, t, and let us calculate the rate at which F varies for a moving particle. This we shall denote by $\dfrac{\partial F}{\partial t}$, the symbol $\dfrac{\partial}{\partial t}$ being used to express a differentiation following the motion of the fluid. At the time $t + dt$ the particle which at the time t was in the position (x, y, z) is in the position $(x + udt, \; y + vdt, \; z + wdt)$, and therefore the corresponding value of F is

$$F + \frac{dF}{dt}\,dt + \frac{dF}{dx}\,udt + \frac{dF}{dy}\,vdt + \frac{dF}{dz}\,wdt.$$

Since the new value of F for the moving particle is also expressed by $F + \dfrac{\partial F}{\partial t}\,dt$, we have

$$\frac{\partial F}{\partial t} = \frac{dF}{dt} + u\frac{dF}{dx} + v\frac{dF}{dy} + w\frac{dF}{dz} \ldots\ldots\ldots \ldots\ldots(1).$$

6. Let p be the pressure, ρ the density, X, Y, Z the components of the external impressed forces per unit mass, at the point (x, y, z) at the time t. Let us take a rectangular element having its centre at (x, y, z), and its edges dx, dy, dz parallel to the co-ordinate axes. The rate at which the x-component of the momentum of this element is increasing is $\rho\,dx\,dy\,dz\,\dfrac{\partial u}{\partial t}$; and this must be equal to the x-component of the forces acting on the element. Of these the external impressed forces give $\rho\,dx\,dy\,dz\,X$. The pressure on the yz-face which is nearest the origin will be ultimately $\left(p - \frac{1}{2}\dfrac{dp}{dx}\,dx\right)dy\,dz$, that on the opposite face $\left(p + \frac{1}{2}\dfrac{dp}{dx}\,dx\right)dy\,dz$. The difference of these gives a resultant $-\dfrac{dp}{dx}\,dx\,dy\,dz$ in the direction of x-positive. The pressures on the remaining faces are perpendicular to x. We have then

$$\rho\,dx\,dy\,dz\,\frac{\partial u}{\partial t} = \rho\,dx\,dy\,dz\,X - \frac{dp}{dx}\,dx\,dy\,dz.$$

Substituting the value of $\frac{\partial u}{\partial t}$ from (1) we obtain

$$\frac{du}{dt} + u\frac{du}{dx} + v\frac{du}{dy} + w\frac{du}{dz} = X - \frac{1}{\rho}\frac{dp}{dx},$$

and, in like manner,

$$\left.\begin{array}{l}\dfrac{dv}{dt} + u\dfrac{dv}{dx} + v\dfrac{dv}{dy} + w\dfrac{dv}{dz} = Y - \dfrac{1}{\rho}\dfrac{dp}{dy}, \\[2mm] \dfrac{dw}{dt} + u\dfrac{dw}{dx} + v\dfrac{dw}{dy} + w\dfrac{dw}{dz} = Z - \dfrac{1}{\rho}\dfrac{dp}{dz}.\end{array}\right\}\dots\dots(2).$$

7. We have thus *three* equations connecting the *five* unknown quantities u, v, w, p, ρ. We require therefore two additional equations. One of these is furnished by a relation between p and ρ, the form of which depends on the physical constitution of the particular fluid which is the subject of investigation. For the case of a gas kept at a uniform temperature we have Boyle's Law

$$p = k\rho \dots\dots\dots\dots\dots\dots\dots (3).$$

If we have a gas in motion of such a nature that we may neglect the loss or gain of heat by an element due to conduction and radiation, the relation is

$$p = k'\rho^{\gamma} \dots\dots\dots\dots\dots\dots\dots(4),$$

where $\gamma = 1{\cdot}41$ for air. In the case of an 'incompressible' fluid, or liquid, we have

$$\rho = \text{constant} \dots\dots\dots\dots\dots\dots\dots(5).$$

8. The remaining equation is a kinematical relation between u, v, w, ρ obtained as follows. If V denote the volume of a moving element of fluid, we have, on account of the constancy of mass,

$$\frac{\partial . \rho V}{\partial t} = 0$$

or

$$\frac{\partial \rho}{\partial t} V + \rho \frac{\partial V}{\partial t} = 0 \dots\dots\dots\dots\dots\dots(6).$$

Now the rate of increase of volume of a moving region is evidently expressed by the surface-integral of the normal velocity outwards, taken all over the boundary. If the region in question be that occupied by the matter which at time t fills the rectangular element of Art. 6, the parts of this surface-integral due to the two yz-faces are

$$\left(u + \tfrac{1}{2}\frac{du}{dx}\,dx\right)dy\,dz, \text{ and } -\left(u - \tfrac{1}{2}\frac{du}{dx}\,dx\right)dy\,dz,$$

which give together $\frac{du}{dx}\,dx\,dy\,dz$. Calculating in the same way the parts due to the other faces, we find

$$\frac{\partial V}{\partial t} = \left(\frac{du}{dx} + \frac{dv}{dy} + \frac{dw}{dz}\right)dx\,dy\,dz.$$

Since we have also $V = dx\,dy\,dz$, (6) becomes

$$\frac{\partial \rho}{\partial t} + \rho\left(\frac{du}{dx} + \frac{dv}{dy} + \frac{dw}{dz}\right) = 0 \ldots\ldots\ldots\ldots\ldots(7),$$

or, as it may be written,

$$\frac{d\rho}{dt} + \frac{d\cdot\rho u}{dx} + \frac{d\cdot\rho v}{dy} + \frac{d\cdot\rho w}{dz} = 0 \ldots\ldots\ldots\ldots\ldots(8).$$

This is called the 'equation of continuity.'

If the fluid be incompressible though not necessarily of uniform density, the value of ρ does not alter as we follow any element, i.e. $\frac{\partial \rho}{\partial t} = 0$, so that (7) becomes

$$\frac{du}{dx} + \frac{dv}{dy} + \frac{dw}{dz} = 0 \ldots\ldots\ldots\ldots\ldots(9).$$

The expression

$$\frac{du}{dx} + \frac{dv}{dy} + \frac{dw}{dz},$$

which, as we have seen, measures the rate of increase of volume of the fluid at the point (x, y, z), is very conveniently termed the 'expansion' at that point.

9. There are certain restrictions as to the values of the dependent variables in the foregoing equations.

Thus u, v, w, p, ρ are essentially single-valued functions.

The quantities u, v, w must be finite, and in general continuous, though we may have isolated surfaces at which the latter restriction does not hold. If the fluid move so as always to form a continuous mass, a certain condition, given in Art. 10, must be satisfied at such a surface.

The quantity p is necessarily continuous, and finite. It is also essentially positive, at all events in the case of ordinary fluids, which cannot sustain more than an infinitesimal amount of *tension* without rupture. Hence if in any of our investigations we be led to negative values of p, the state of motion given by the formulæ is an impossible one. At the moment when, according to the formulæ, p would change from positive through zero to negative, either the fluid parts asunder, or a surface of discontinuity is formed, so that the conditions of the problem are entirely changed. See Art. 94.

The quantity ρ is finite and positive, but not necessarily continuous.

10. The equations, which have been obtained so far, relate to the interior of the fluid. Besides these we have, in general, to satisfy certain boundary conditions, the nature of which varies according to the circumstances of the case.

Let
$$F(x, y, z, t) = 0 \dots\dots\dots\dots\dots (10)$$
be the equation to a surface bounding the fluid. The velocity relative to this surface of a particle lying in it must be wholly tangential (or else zero), for otherwise we should have a finite flow of liquid across the surface, which contradicts the assumption that the latter is a boundary. The instantaneous rate of variation of F for a surface-particle must therefore be zero, *i.e.* we have

$$\frac{\partial F}{\partial t} = 0 \dots\dots\dots\dots\dots (11).$$

This must hold at every point of the surface represented by (10).

At a *fixed* boundary we have $\dfrac{dF}{dt} = 0$, so that (11) becomes

$$u\frac{dF}{dx} + v\frac{dF}{dy} + w\frac{dF}{dz} = 0,$$

or, if l, m, n be the direction-cosines of the normal to the surface,
$$lu + mv + nw = 0 \dots\dots\dots\dots\dots(12).$$

If $F = 0$ be the equation of a surface of discontinuity, *i.e.* a surface such that the values of u, v, w change abruptly as we pass from one side to the other, we have

$$\frac{dF}{dt} + u_1\frac{dF}{dx} + v_1\frac{dF}{dy} + w_1\frac{dF}{dz} = 0,$$

and
$$\frac{dF}{dt} + u_2\frac{dF}{dx} + v_2\frac{dF}{dy} + w_2\frac{dF}{dz} = 0,$$

where the suffixes are used to distinguish the two sides of the surface. By subtraction we find

$$l\,(u_1 - u_2) + m\,(v_1 - v_2) + n\,(w_1 - w_2) = 0 \ldots \ldots (13).$$

The same relation holds at the common surface of two different fluids in contact; and also, since in the proof of (11) no assumption is made as to the nature of the medium of which (10) is a boundary, at the common surface of a fluid and a moving solid.

The truth of (13), of which (12) is a particular case, is otherwise obvious from the consideration that the velocity normal to the surface must be, in each of the cases mentioned, the same on both sides.

11. The equation (11) expresses the condition that if the motion be continuous the particles which at any instant lie in the bounding surface lie in it always. For (11) expresses that no fluid crosses the surface $F = 0$; and the same thing necessarily holds of every surface which moves so as to consist always of the same series of particles. If then we draw a surface parallel and infinitely close to $F = 0$, and suppose it to move with the particles of which it is composed, the stratum of fluid which is included between this and $F = 0$, and which in virtue of the continuity of the motion remains always infinitely thin, must always consist of the same matter; whence the truth of the above statement.

It has been suggested that (11) would be satisfied if the particles of fluid were to move relatively to the surface $F = 0$ in paths touching it each at one point only. The above considerations shew that this is not possible for a system of material particles moving in a continuous manner; although it would be so for mere geometrical points which might coincide with and pass through one another*. It is, indeed, difficult to understand how, in the case supposed, the particles which are receding from the surface are to keep clear of those which are approaching it.

12. In the above method of establishing the fundamental equations we calculate the rate of change of the properties of a definite

* The student may take as an illustration the motion of a series of points given by the formulæ

$$u = \pm x, \quad v = c, \quad w = 0,$$

the upper sign in u being taken for points receding from the fixed boundary $x = 0$, the lower for points approaching it.

portion of matter as it moves along.   In another method, which is indeed more consistent with the Eulerian notation, we fix our attention on a certain region of space, and investigate the change in its properties produced as well by the flow of matter inwards and outwards across the boundary as by the action of external forces on the included mass*.

Let Q denote the measure, estimated per unit volume, of any quantity connected with the properties of a fluid, and let us calculate the rate of increase of Q in a rectangular space $dx\,dy\,dz$ having its centre at (x, y, z).   This is expressed by

$$\frac{dQ}{dt}\,dx\,dy\,dz \dots\dots\dots\dots\dots\dots(14).$$

·Now the amount of Q which enters per unit time the specified region across the yz-face nearest the origin is $\left(Qu - \tfrac{1}{2}\dfrac{d\,.\,Qu}{dx}\,dx\right)dy\,dz$, and the amount which leaves the region in the same time by the opposite face is $\left(Qu + \tfrac{1}{2}\dfrac{d\,.\,Qu}{dx}\,dx\right)dy\,dz$.   The two faces together give a gain of $-\dfrac{d\,.\,Qu}{dx}\,dx\,dy\,dz$ per unit time.   Calculating in the same way the effect of the flow across the remaining faces, we have for the total gain of Q due to the flow across the boundary the formula

$$-\left(\frac{d\,.\,Qu}{dx}+\frac{d\,.\,Qv}{dy}+\frac{d\,.\,Qw}{dz}\right)dx\,dy\,dz\dots\dots\dots\dots(15).$$

First, let us consider the change of mass, *i.e.* we put $Q=\rho$, the mass per unit volume.   Since the quantity of matter in any region can vary only in consequence of the flow across the boundary, the expressions (14) and (15) must in this case be equal; this gives the equation of continuity in the form (8).

Next, let us take the change of momentum, making $Q=\rho u$, the momentum parallel to x per unit mass.   The momentum contained in the space $dx\,dy\,dz$ is affected not only by the passage of matter carrying its momentum with it across the boundary, but also by the forces acting on the included matter, viz. the pressure and the external impressed forces.   The effect of these resolved

* Seo Maxwell, On the Dynamical Theory of Gases, *Phil. Trans.* 1867, p. 71. Also, Greenhill, *Solutions of Cambridge Problems for* 1875, p. 178.

parallel to x is found as in Art. 6, to be

$$\left(\rho X - \frac{dp}{dx}\right) dx\,dy\,dz\dots\dots\dots\dots\dots(16).$$

Hence, (14) is now equal to (15) and (16) combined, which gives

$$\frac{d\cdot\rho u}{dt} + \frac{d\cdot\rho u^2}{dx} + \frac{d\cdot\rho uv}{dy} + \frac{d\cdot\rho uw}{dz} = \rho X - \frac{dp}{dx}\dots\dots(17).$$

Performing the differentiations, and simplifying by means of the equation of continuity, we are led again to the first of equations (2), and in like manner the second and third equations may be obtained.

13. Another interesting application of the method of Art. 12 is to make $Q = (\frac{1}{2}q^2 + V + E)\rho$, the energy per unit mass. Here q denotes the resultant velocity $\sqrt{(u^2 + v^2 + w^2)}$, V the potential energy per unit mass with reference to the external impressed forces $\left(\text{viz. we have } X = -\frac{dV}{dx}, \&c.\right)$, and E the intrinsic energy. In a liquid we have $E = 0$. If the system of external forces do not change with the time the alteration in the energy contained within the space $dx\,dy\,dz$ is due to the flow of matter carrying its energy with it, and to the work done on the contained matter by the pressure of the surrounding fluid. The total rate at which this pressure works is

$$-\left(\frac{d\cdot pu}{dx} + \frac{d\cdot pv}{dy} + \frac{d\cdot pw}{dz}\right) dx\,dy\,dz\dots\dots\dots(18).$$

The verification of the formula obtained by equating (14) to the sum of (15) and (18) is left as an exercise for the student.

14. To obtain by the same method a proof of the surface-condition (11) of Art. 10, let in Fig. 1 (Art. 3) P denote a point of the fluid infinitely close to the surface $F = 0$; and let A, B, C be the points in which this surface is met by three straight lines drawn through P parallel to the axes of co-ordinates. Then if $PA, PB, PC = \alpha, \beta, \gamma$ respectively, we have

$$\alpha = -F_0 \div \frac{dF}{dx}, \&c.,$$

where F_0 denotes the value of the function F at $P(x, y, z)$. The rate of flow of matter into the space included between the three planes meeting in P, and the surface $F = 0$, is ultimately

$$\tfrac{1}{2}\rho\,(u\beta\gamma + v\gamma\alpha + w\alpha\beta);$$

and the rate of increase of the mass included in this space is ultimately $\dfrac{d}{dt}\,(\tfrac{1}{6}\rho\alpha\beta\gamma)$. Equating these expressions, substituting for α, β, γ their values, and omitting infinitesimals of higher order than the second, we readily find

$$u\frac{dF}{dx} + v\frac{dF}{dy} + w\frac{dF}{dz} = -\frac{dF}{dt},$$

which agrees with (11).

Impulsive Generation of Motion.

15. If at any instant impulsive forces act on the mass of the fluid, or if the boundary conditions suddenly change, a sudden alteration in the motion may take place. The latter case may arise, for instance, when a solid immersed in the fluid is suddenly set in motion.

Let ρ be the density, u, v, w the component velocities immediately before, u', v', w' those immediately after the impulse, X', Y', Z' the components of the external impulsive forces per unit mass, ϖ the impulsive pressure, at the point (x, y, z). The change of momentum parallel to x of the element defined in Art. 6 is then $\rho\,dx\,dy\,dz\,(u' - u)$; the x-component of the external impulsive forces is $\rho\,dx\,dy\,dz\,X'$, and the resultant impulsive pressure in the same direction is $-\dfrac{d\varpi}{dx}\,dx\,dy\,dz$. Since an impulse is to be regarded as an infinitely great force acting for an infinitely short time (τ, say), the effects of all finite forces during this interval are neglected.

Hence,

$$\rho\,dx\,dy\,dz\,(u' - u) = \rho\,dx\,dy\,dz\,X' - \frac{d\varpi}{dx}\,dx\,dy\,dz,$$

or

$$u' - u = X' - \frac{1}{\rho}\frac{d\varpi}{dx}.$$

Similarly,

$$v' - v = Y' - \frac{1}{\rho}\frac{d\varpi}{dy},$$

$$w' - w = Z' - \frac{1}{\rho}\frac{d\varpi}{dz}.$$

$$\left.\right\}\dots\dots\dots(19).$$

These equations might also have been deduced from (2), by multiplying the latter by dt, integrating between the limits 0 and τ, putting $X' = \int_0^\tau X\,dt$, &c., $\varpi = \int_0^\tau p\,dt$, and then making τ vanish.

In a gas an infinite pressure would involve an infinite density; whereas no change of density can occur during the infinitely short time τ of the impulse. Hence, in applying (19) to the case of a gas we must put $\varpi = 0$, whence

$$u' - u = X', \quad v' - v = Y', \quad w' - w = Z' \dots\dots\dots\dots\dots(20).$$

In a liquid, on the other hand, an instantaneous change of motion can be produced by the action of impulsive pressures only, even when no impulsive forces act bodily on the mass. In this case we have X', Y', Z' each $= 0$, so that

$$\left.\begin{aligned} u' - u &= -\frac{1}{\rho}\frac{d\varpi}{dx}, \\ v' - v &= -\frac{1}{\rho}\frac{d\varpi}{dy}, \\ w' - w &= -\frac{1}{\rho}\frac{d\varpi}{dz}. \end{aligned}\right\} \dots\dots\dots\dots\dots\dots(21).$$

If we differentiate these equations with respect to x, y, z, respectively, and add, and if we further suppose the density to be uniform, we find by (9) that

$$\frac{d^2\varpi}{dx^2} + \frac{d^2\varpi}{dy^2} + \frac{d^2\varpi}{dz^2} = 0.$$

The problem then, in any given case, is to determine a value of ϖ satisfying this equation and the proper boundary conditions*; the instantaneous change of motion is then given by (21).

The Lagrangian Forms of the Equations.

16.   Let a, b, c be the initial co-ordinates of any particle of fluid, x, y, z its co-ordinates at time t. We here consider x, y, z as functions of the independent variables a, b, c, t; their values in terms of these quantities give the whole history of every particle of the fluid. The velocities parallel to the axes of co-ordinates of

---

* It will appear in Chapter III. that (save as to additive constants) there is only one value of ϖ which does this.

the particle (a, b, c) at time t are $\dfrac{dx}{dt}, \dfrac{dy}{dt}, \dfrac{dz}{dt}$, and the component accelerations in the same directions are $\dfrac{d^2x}{dt^2}, \dfrac{d^2y}{dt^2}, \dfrac{d^2z}{dt^2}$. Let p and ρ be the pressure and density in the neighbourhood of this particle at time t; X, Y, Z the components of the external impressed forces per unit mass acting there. Considering the motion of the mass of fluid which at time t occupies the differential element of volume $dx\,dy\,dz$, we find by the same reasoning as in Art. 6,

$$\left.\begin{aligned}
\frac{d^2x}{dt^2} &= X - \frac{1}{\rho}\frac{dp}{dx}, \\
\frac{d^2y}{dt^2} &= Y - \frac{1}{\rho}\frac{dp}{dy}, \\
\frac{d^2z}{dt^2} &= Z - \frac{1}{\rho}\frac{dp}{dz}.
\end{aligned}\right\}$$

These equations contain differential coefficients with respect to x, y, z, whereas our independent variables are a, b, c, t. To eliminate these differential coefficients, we multiply the above equations by $\dfrac{dx}{da}, \dfrac{dy}{da}, \dfrac{dz}{da}$, respectively, and add; a second time by $\dfrac{dx}{db}, \dfrac{dy}{db}, \dfrac{dz}{db}$, and add; and again a third time by $\dfrac{dx}{dc}, \dfrac{dy}{dc}, \dfrac{dz}{dc}$, and add. We thus get the three equations

$$\left.\begin{aligned}
\left(\frac{d^2x}{dt^2} - X\right)\frac{dx}{da} + \left(\frac{d^2y}{dt^2} - Y\right)\frac{dy}{da} + \left(\frac{d^2z}{dt^2} - Z\right)\frac{dz}{da} + \frac{1}{\rho}\frac{dp}{da} &= 0, \\
\left(\frac{d^2x}{dt^2} - X\right)\frac{dx}{db} + \left(\frac{d^2y}{dt^2} - Y\right)\frac{dy}{db} + \left(\frac{d^2z}{dt^2} - Z\right)\frac{dz}{db} + \frac{1}{\rho}\frac{dp}{db} &= 0, \\
\left(\frac{d^2x}{dt^2} - X\right)\frac{dx}{dc} + \left(\frac{d^2y}{dt^2} - Y\right)\frac{dy}{dc} + \left(\frac{d^2z}{dt^2} - Z\right)\frac{dz}{dc} + \frac{1}{\rho}\frac{dp}{dc} &= 0.
\end{aligned}\right\}\dots(22).$$

These are the Lagrangian forms of the dynamical equations.

17. As before, two additional equations are required. We have, first, a relation between p and ρ of the form (3), (4), or (5), as the case may be. To find the form which the equation of continuity assumes in terms of our present variables, we consider the element of fluid which originally occupied a rectangular parallelepiped having the corner nearest the origin at the point (a, b, c), and its edges da, db, dc parallel to the axes. At the time t the

same element forms an oblique parallelepiped. The corner corresponding to (a, b, c) has for its co-ordinates x, y, z; and the co-ordinates relative to this point of the other extremities of the three edges meeting in it are respectively $\frac{dx}{da} da$, $\frac{dy}{da} da$, $\frac{dz}{da} da$;

$\frac{dx}{db} db$, $\frac{dy}{db} db$, $\frac{dz}{db} db$; $\frac{dx}{dc} dc$, $\frac{dy}{dc} dc$, $\frac{dz}{dc} dc$. The volume of the parallelepiped is therefore *

$$\begin{vmatrix} \dfrac{dx}{da}, & \dfrac{dy}{da}, & \dfrac{dz}{da} \\[2mm] \dfrac{dx}{db}, & \dfrac{dy}{db}, & \dfrac{dz}{db} \\[2mm] \dfrac{dx}{dc}, & \dfrac{dy}{dc}, & \dfrac{dz}{dc} \end{vmatrix} da\, db\, dc,$$

or, as it is often written,

$$\frac{d(x, y, z)}{d(a, b, c)} da\, db\, dc.$$

Hence, since the mass of the element is unchanged, we have

$$\rho \frac{d(x, y, z)}{d(a, b, c)} = \rho_0 \quad \dots\dots\dots\dots\dots (23),$$

where ρ_0 is the initial density at (a, b, c).

In the case of an incompressible fluid $\rho = \rho_0$, so that (23) becomes

$$\frac{d(x, y, z)}{d(a, b, c)} = 1 \quad \dots\dots\dots\dots\dots (24).$$

Weber's Transformation.

18. If the forces X, Y, Z have a potential, i.e. if they can be expressed as the partial differential coefficients with respect to x, y, z of a single function which we denote by $-V$ (so that V is the potential energy, due to those forces, of unit mass placed in the position (x, y, z)), the equations (22) may be written

$$\frac{d^2x}{dt^2}\frac{dx}{da} + \frac{d^2y}{dt^2}\frac{dy}{da} + \frac{d^2z}{dt^2}\frac{dz}{da} = -\frac{dV}{da} - \frac{1}{\rho}\frac{dp}{da},$$

&c., &c.

---

* Salmon, *Geometry of Three Dimensions.*

Let us integrate these equations with respect to t between the limits 0 and t. We remark that

$$\int_0^t \frac{d^2x}{dt^2}\frac{dx}{da}\,dt = \left[\frac{dx}{dt}\frac{dx}{da}\right]_0^t - \int_0^t \frac{dx}{dt}\frac{d^2x}{da\,dt}\,dt$$

$$= \frac{dx}{dt}\frac{dx}{da} - u_0 - \tfrac{1}{2}\frac{d}{da}\int_0^t \left(\frac{dx}{dt}\right)^2 dt,$$

where u_0 is the initial value of the x-component of velocity of the particle (a, b, c). Hence if we write

$$\chi = \int_0^t \left[-V - \int\frac{dp}{\rho} + \tfrac{1}{2}\left\{\left(\frac{dx}{dt}\right)^2 + \left(\frac{dy}{dt}\right)^2 + \left(\frac{dz}{dt}\right)^2\right\}\right]dt \ldots(25),$$

we have

$$\left.\begin{array}{l}
\dfrac{dx}{dt}\dfrac{dx}{da} + \dfrac{dy}{dt}\dfrac{dy}{da} + \dfrac{dz}{dt}\dfrac{dz}{da} - u_0 = \dfrac{d\chi}{da}; \\[2mm]
\text{and, similarly,} \quad \dfrac{dx}{dt}\dfrac{dx}{db} + \dfrac{dy}{dt}\dfrac{dy}{db} + \dfrac{dz}{dt}\dfrac{dz}{db} - v_0 = \dfrac{d\chi}{db}, \\[2mm]
\dfrac{dx}{dt}\dfrac{dx}{dc} + \dfrac{dy}{dt}\dfrac{dy}{dc} + \dfrac{dz}{dt}\dfrac{dz}{dc} - w_0 = \dfrac{d\chi}{dc}.
\end{array}\right\} \ldots\ldots\ldots(26)^*.$$

These three equations, together with

$$\frac{d\chi}{dt} = -V - \int\frac{dp}{\rho} + \tfrac{1}{2}\left\{\left(\frac{dx}{dt}\right)^2 + \left(\frac{dy}{dt}\right)^2 + \left(\frac{dz}{dt}\right)^2\right\} \ldots\ldots(27),$$

and the equation of continuity, are the partial differential equations to be satisfied by the five unknown quantities x, y, z, p, χ; ρ being supposed already eliminated by means of one of the relations of Art. 7.

In the case of a liquid, p occurs in (27) only, so that (26) and (24) may be employed to find x, y, z, and χ, while p may be found afterwards from (27).

The initial conditions to be satisfied are $x = a$, $y = b$, $z = c$, $\chi = 0$. The boundary conditions vary with the particular problem under investigation.

19. The equations (26) and (27) may be applied to find the equations of impulsive motion of a liquid. Let the impulse act from $t = 0$ to $t = \tau$, where τ is infinitely small, and let τ be the

---

* H. Weber, *Crelle*, t. 68.

upper limit of integration in (25). We find $\chi = -V'' - \dfrac{\varpi}{\rho}$, where ϖ is the impulsive pressure and V' the potential of the external impulsive forces at the point (a, b, c). Since $x = a$, $y = b$, $z = c$, we have, by (26),

$$\frac{dx}{dt} - u_0 = -\frac{dV'}{da} - \frac{1}{\rho}\frac{d\varpi}{da}, \quad \&c., \&c.,$$

which agree with the equations of Art. 15.

20. In the method of Art. 16 the quantities a, b, c need not be restricted to mean the initial co-ordinates of a particle; they may be considered to be any three quantities which serve to identify a particle, and which vary continuously from one particle to another. If we thus generalize the meanings of a, b, c, the form of equations (22) is not altered; to find the form which (23) assumes, let x_0, y_0, z_0 now denote the initial co-ordinates of the particle to which a, b, c refer. The initial volume of the parallelepiped, three of whose edges are drawn from the particle (a, b, c) to the particles $(a + da, b, c)$, $(a, b + db, c)$, $(a, b, c + dc)$, respectively, is

$$\frac{d(x_0, y_0, z_0)}{d(a, b, c)}\, da\, db\, dc,$$

so that instead of (23) we have

$$\rho\,\frac{d(x, y, z)}{d(a, b, c)} = \rho_0\frac{d(x_0, y_0, z_0)}{d(a, b, c)} \quad\ldots\ldots\ldots\ldots(28),$$

and for incompressible fluids

$$\frac{d(x, y, z)}{d(a, b, c)} = \frac{d(x_0, y_0, z_0)}{d(a, b, c)} \quad\ldots\ldots\ldots\ldots\ldots(29).$$

21. If we compare the two forms of the fundamental equations to which we have been led, we notice that the Eulerian equations of motion are linear and of the first order, whilst the Lagrangian equations are of the second order, and also contain products of differential coefficients. In Weber's transformation the latter are replaced by a system of equations of the first order, and of the second degree. The Eulerian equation of continuity is also much simpler than the Lagrangian, especially in the case of liquids. In these respects, therefore, the Eulerian forms of the equations possess great advantages over the Lagrangian. Again, the form in

which the solution of the Eulerian equations appears corresponds, in many cases, more nearly to what we wish to know as to the motion of a fluid, our object being, in general, to gain a knowledge of the state of motion of the fluid mass at any instant, rather than to trace the career of individual particles.

* On the other hand, whenever the fluid is bounded by a moving surface, the Lagrangian method possesses certain theoretical advantages. In the Eulerian method the functions u, v, w have no existence beyond this surface, and hence the range of values of x, y, z for which these functions exist varies in consequence of the motion which we have to investigate. In the other method, on the contrary, the range of values of the independent variables a, b, c is given once for all by the initial conditions.

The difficulty, however, of integrating the Lagrangian equations has hitherto prevented their application except in certain very special cases. Accordingly in this treatise we deal almost exclusively with the Eulerian equations. The integration and simplification of these in certain cases form the subject of the following chapter.

* H. Weber, *Crelle*, t. 68.

CHAPTER II.

22. In most cases of interest the external impressed forces have a potential; viz. we have

$$X = -\frac{dV}{dx}, \quad Y = -\frac{dV}{dy}, \quad Z = -\frac{dV}{dz} \dots\dots\dots\dots(1).$$

In a large and important class of cases the component velocities u, v, w can be similarly expressed as the partial differential coefficients of a function ϕ, so that

$$u = \frac{d\phi}{dx}, \quad v = \frac{d\phi}{dy}, \quad w = \frac{d\phi}{dz} \dots\dots\dots\dots(2).$$

Such a function is called a 'velocity-potential,' from its analogy to the potential function which occurs in·the theories of Attractions, Electrostatics, &c. The general theory of the velocity-potential is reserved for the next chapter; but we give at once a proof of the following important theorem :

23. If a velocity-potential exist, at any one instant, for any finite portion of a perfect fluid in motion under the action of forces which have a potential, then, provided the density of the fluid be either constant or a function of the pressure only, a velocity-potential exists for the same portion of the fluid at all subsequent instants.

In the equations of Art. 18, let the instant at which the velocity-potential ϕ_0 exists be taken as the origin of time; we have then

$$u_0 da + v_0 db + w_0 dc = d\phi_0,$$

throughout the portion of the mass in question. Multiplying the three equations (26) Art. 18 in order by da, db, dc, and adding, we get

$$\frac{dx}{dt}\,dx + \frac{dy}{dt}\,dy + \frac{dz}{dt}\,dz - (u_0 da + v_0 db + w_0 dc) = d\chi.$$

or, with our present notation,

$$udx + vdy + wdz = d\,(\phi_0 + \chi) = d\phi, \text{ say};$$

which proves the theorem.

It is to be particularly noticed that this continued existence of a velocity-potential is predicated, not of regions of space, but of portions of matter. A portion of matter for which a velocity-potential exists moves about and carries this property with it, but the portion of space which it originally occupied may, in the course of the motion, come to be occupied by matter which did not originally possess this property, and which therefore cannot have acquired it.

The above theorem, stated in an imperfect form by Lagrange in Section XI. of the *Mécanique Analytique*, was first placed in its proper light by Cauchy. Other proofs, to be reproduced further on, have since been given by Stokes*, Helmholtz, and Thomson. A careful criticism of Lagrange's and other proofs has been given by Stokes*.

24. The class of cases in which a velocity-potential exists includes all those where the motion has originated from rest under the action of forces of the kind here supposed; for then we have, initially,

$$udx + vdy + wdz = 0,$$

or $$\phi = \text{const.}$$

Again, if the motion be so slow that the squares and products of u, v, w and their first differential coefficients may be neglected, the equations (2) become

$$\frac{du}{dt} = -\frac{dV}{dx} - \frac{1}{\rho}\frac{dp}{dx}, \ \&c., \ \&c.;$$

so that $$\frac{du}{dt}\,dx + \frac{dv}{dt}\,dy + \frac{dw}{dt}\,dz$$

* *Camb. Phil. Trans.* Vol. VIII. (1845), p. 305 et seq.

is an exact differential.   Hence, integrating, we see that

$$ud x + v dy + w dz$$

consists of two parts, one of which is an exact differential, whilst the other does not contain t.   In some cases, for example, when the motion is wholly periodic, we can assert that the latter part is zero, and therefore, that a velocity-potential exists.

25.   Under the circumstances stated in Art. 23, the equations of Art. 6 are at once integrable throughout that portion of the mass for which a velocity-potential exists.   For, in virtue of the relations $\frac{dv}{dz} = \frac{dw}{dy}$, $\frac{dw}{dx} = \frac{du}{dz}$, $\frac{du}{dy} = \frac{dv}{dx}$, which are implied in (2), the equations in question may be written

$$\frac{d^2\phi}{dx dt} + u\frac{du}{dx} + v\frac{dv}{dx} + w\frac{dw}{dx} = -\frac{dV}{dx} - \frac{1}{\rho}\frac{dp}{dx}, \text{ &c. &c.}$$

These have the common integral

$$\frac{d\phi}{dt} + \tfrac{1}{2}q^2 = -V - \int\frac{dp}{\rho} + F(t)\dots\dots\dots\dots\dots(3).$$

Here q denotes the resultant velocity $\sqrt{(u^2 + v^2 + w^2)}$, and $F(t)$ is an arbitrary function of t, which may however be supposed included in $\frac{d\phi}{dt}$, since, by (2), the values of u, v, w are not thereby affected.

For incompressible fluids the equation (3) becomes

$$\frac{d\phi}{dt} + \tfrac{1}{2}q^2 = -V - \frac{p}{\rho} + F(t)\dots\dots\dots\dots\dots(4),$$

whilst the equation of continuity ((9) of Art. 8) assumes the form

$$\frac{d^2\phi}{dx^2} + \frac{d^2\phi}{dy^2} + \frac{d^2\phi}{dz^2} = 0\dots\dots\dots\dots\dots(5).$$

In any problem to which these equations apply, and where the boundary-conditions are purely kinematical, the process of solution is as follows.   We must first find a function ϕ satisfying (5) and the given boundary-conditions; then substituting in (4) we get the value of p.   Since the latter equation contains an arbitrary function of t, the complete determination of p requires a knowledge of its value at some point of the fluid for all values of t.

26.   A comparison of equations (2) with the equations of Art. 15 gives a simple physical interpretation of the velocity-potential.

Any actual state of motion of a liquid, for which a velocity-potential exists, could be produced instantaneously from rest by the application of a properly chosen system of impulsive pressures. This is evident from equations (21) Art. 15, which shew, moreover, that

$\phi = -\dfrac{\varpi}{\rho} + \text{const.}$; so that $\varpi = C - \rho\phi$ gives the requisite system. In the same way $\varpi = \rho\phi + C$ gives the system of impulsive pressures which would completely stop the motion. The occurrence of an arbitrary constant in these expressions shews, what is otherwise evident, that a pressure uniform throughout a liquid mass produces no effect on its motion.

In the case of a gas, ϕ is the potential of the external impulsive forces by which the actual motion at any instant could be produced instantaneously from rest.

A state of motion for which a velocity-potential does not exist cannot be generated or destroyed by the action of impulsive pressures, or of external impulsive forces having a potential.

27.   The existence of a velocity-potential indicates, besides, certain *kinematical* properties of the motion.

A 'line of motion' is defined to be a line drawn from point to point, so that its direction is everywhere that of the motion of the fluid. The differential equations of the system of such lines are

$$\frac{dx}{u} = \frac{dy}{v} = \frac{dz}{w} \dots\dots\dots\dots\dots\dots\dots\dots(6).$$

The relations (2) shew that when a velocity-potential exists the lines of motion are everywhere perpendicular to a series of surfaces, viz. the surfaces $\phi = \text{const.}$ These are called the surfaces of equal velocity-potential, or more shortly, the equipotential surfaces.

Again, if from the point (x, y, z) we draw a linear element ds in the direction (l, m, n), the velocity resolved in this direction is $lu + mv + nw$, or $\dfrac{d\phi}{dx}\dfrac{dx}{ds} + \dfrac{d\phi}{dy}\dfrac{dy}{ds} + \dfrac{d\phi}{dz}\dfrac{dz}{ds}$, which $= \dfrac{d\phi}{ds}$. The velocity in any direction is therefore equal to the rate of increase of ϕ in that direction.

Taking ds in the direction of the normal to the surface $\phi = $ const. we see that if a series of such surfaces be drawn so that the difference between the values of ϕ for two consecutive surfaces is constant and infinitely small, the velocity at any point will be inversely proportional to the distance between two consecutive surfaces in the neighbourhood of that point.

Hence, if any equipotential surface intersect itself, the velocity is zero at every point of the intersection.

The intersection of two distinct equipotential surfaces would imply an infinite velocity at all points of the intersection.

Steady Motion.

28.   When at every point the velocity at that point is constant in magnitude and direction, *i.e.* when

$$\frac{du}{dt} = 0, \quad \frac{dv}{dt} = 0, \quad \frac{dw}{dt} = 0 \dots\dots\dots\dots\dots(7)$$

everywhere, the motion is said to be 'steady.'

In steady motion the lines of motion coincide with the paths of the particles and are in this case called 'stream-lines.' For let P, Q be two consecutive points on a line of motion. A particle which is at any instant at P is moving in the direction of the tangent at P, and will, therefore, after an infinitely short time arrive at Q. The motion being steady, the lines of motion remain the same. Hence the direction of motion at Q is along the tangent to the same line of motion, *i.e.* the particle continues to describe the line of motion.

In steady motion the equation (3) becomes

$$\int \frac{dp}{\rho} = - V - \tfrac{1}{2}q^2 + \text{constant} \dots\dots\dots\dots\dots(8).$$

The equations of motion may however in this case be integrated to a certain extent without assuming the existence of a velocity-potential. For if ds denote an element of a stream-line, we have $u = q\dfrac{dx}{ds}$, &c. Substituting in the equations of motion we have, remembering (7),

$$q\frac{du}{ds} = X - \frac{1}{\rho}\frac{dp}{dx},$$

with two similar equations. Multiplying these in order by $\dfrac{dx}{ds}$, $\dfrac{dy}{ds}$, $\dfrac{dz}{ds}$, and adding, we have

$$u\frac{du}{ds} + v\frac{dv}{ds} + w\frac{dw}{ds} = -\frac{dV}{ds} - \frac{1}{\rho}\frac{dp}{ds},$$

or, integrating along the stream-line,

$$\int\frac{dp}{\rho} = -V - \tfrac{1}{2}q^2 + C \quad\ldots\ldots\ldots\ldots\ldots(9).$$

This is of the same form as (8), but is more general in that it does not involve the assumption of the existence of a velocity-potential. It must however be carefully noticed that the 'constant' of equation (8) and the 'C' of equation (9) have very different meanings, the former being an absolute constant, while the latter is constant along any particular stream-line, but may vary as we pass from one stream-line to another.

29. The formula (9) may be deduced from the principle of energy without employing the equations of motion at all. Taking first the particular case of a liquid, let us consider the portion of an infinitely narrow tube, whose walls are formed of stream-lines, included between two cross sections A and B, the direction of motion being from A to B. Let p be the pressure, q the velocity, V the potential of the external forces, σ the area of the cross section at A, and let the values of the same quantities at B be distinguished by accents. In each unit of time a mass $\rho q\sigma$ at A enters the portion of the tube considered, whilst an equal mass $pq'\sigma'$ leaves it at B. Hence $q\sigma = q'\sigma'$. Again, the work done on the mass entering at A is $pq\sigma$ per unit time, whilst the loss of work at B is $p'q'\sigma'$. The former mass brings with it the energy $\rho q\sigma\left(\tfrac{1}{2}q^2 + V\right)$, whilst the latter carries off energy to the amount $\rho q'\sigma'\left(\tfrac{1}{2}q'^2 + V'\right)$. The motion being steady, the portion of the tube considered neither gains nor loses energy on the whole, so that

$$pq\sigma + \rho q\sigma\left(\tfrac{1}{2}q^2 + V\right) = p'q'\sigma' + \rho q'\sigma'\left(\tfrac{1}{2}q'^2 + V'\right).$$

Dividing by $\rho q\sigma \,(= \rho q'\sigma')$, we have

$$\frac{p}{\rho} + \tfrac{1}{2}q^2 + V = \frac{p'}{\rho} + \tfrac{1}{2}q'^2 + V',$$

or, using C in the same sense as before,

$$\frac{p}{\rho} + \tfrac{1}{2}q^2 + V = C \quad\ldots\ldots\ldots\ldots\ldots(10),$$

which is what the equation (9) becomes when ρ is constant.

To prove the corresponding formula for compressible fluids, we remark that the fluid entering at A now brings with it, in addition to its energies of motion and position, the intrinsic energy $-\int pd\left(\frac{1}{\rho}\right)^{*}$, or $-\frac{p}{\rho} + \int\frac{dp}{\rho}$, per unit mass. The addition of these terms converts the equation (10) into the equation (9).

In most cases of motion of gases, the relation (4) of Art. 7 holds, and (9) then becomes

$$\frac{\gamma}{\gamma - 1}\frac{p}{\rho} = -V - \tfrac{1}{2}q^2 + C \ldots\ldots\ldots\ldots\ldots(11).$$

30. Equations (10) and (11) shew that, in steady motion for points along any one stream-line$\dagger$, the pressure is, *cæteris paribus*, greatest where the velocity is least, and *vice versa*. This statement, though opposed to popular notions, is obvious if we reflect that a particle passing from a place of higher to one of lower pressure must have its motion accelerated, and *vice versa*. Some interesting practical illustrations and applications of the principle are given by Mr Froude in *Nature*, Vol. XIII. 1875.

It follows that in any case to which the aforesaid equations apply there is a limit which the velocity cannot exceed if the motion be continuous. For instance, let us suppose that we have a liquid flowing from a reservoir where the motion is sensibly zero, and the pressure equal to P, and that we may neglect the external impressed forces. We have then in (10) $C = \dfrac{P}{\rho}$, and therefore

$$\frac{p}{\rho} = \frac{P}{\rho} - \tfrac{1}{2}q^2 \ldots\ldots\ldots\ldots\ldots(12).$$

Hence if q^2 exceed $\dfrac{2P}{\rho}$, p becomes negative, whereas we know that actual fluids are unable to support more than a very slight, if any,

---

* Tait's *Thermodynamics*, Art. 174 (first edition).

$\dagger$ This restriction is, by (9), unnecessary when a velocity-potential exists.

degree of tension, without rupture. The limiting velocity is, by (12), approximately that with which the fluid would escape from the reservoir into a vacuum. In the case of water at the atmospheric pressure, this velocity is that due to the height of the water-barometer, or roughly, about 45 feet per second.

The question as to what takes place when the limiting velocity is reached will be considered in Art. 94.

31. We conclude this chapter with a few simple applications of the equations.

Example 1. Steady motion under the action of gravity. A vessel is kept filled up to a constant level with liquid which escapes from a small orifice in its walls.

The origin being taken in the upper surface, let the axis of z be vertical, and its positive direction downwards, so that $V = -gz$. If we suppose the area of the upper surface large compared with that of the orifice, the velocity at the former may be neglected. Hence, determining the constant in (10) so that $p = P$ (the atmospheric pressure), when $z = 0$, we have

$$\frac{p}{\rho} = \frac{P}{\rho} + gz - \tfrac{1}{2}q^2 \dots\dots\dots\dots\dots(13).$$

At the surface of the issuing jet we have $p = P$, and therefore

$$q^2 = 2gz \dots\dots\dots\dots\dots\dots(14),$$

i.e. the velocity is that due to the depth below the upper surface. This is *Torricelli's Theorem*.

We cannot however at once apply this result to calculate the rate of efflux of the fluid, for two reasons. In the first place, the issuing fluid must be regarded as made up of a great number of elementary streams converging from all sides towards the orifice. Its motion is not, therefore, throughout the area of the orifice, everywhere perpendicular to this area, but becomes more and more oblique as we pass from the centre of the orifice to the sides. Again, the converging motion of the elementary streams must make the pressure at the orifice somewhat greater in the interior of the jet than at its surface, where it is equal to the atmospheric pressure. The velocity, therefore, in the interior of the jet will be somewhat less than that given by (14).

Experiment shews however that the converging motion above spoken of ceases at a short distance beyond the orifice, and that the jet then becomes approximately cylindrical.

The ratio of the area of the section S' of the jet at this point (called the 'vena contracta') to the area S of the orifice is called the 'coefficient of contraction.' If the orifice be simply a hole in a thin wall, this coefficient is found to be about ·62. If a short cylindrical tube be attached externally, the value of the coefficient is considerably increased; if, on the other hand, there be attached a short tube projecting *inwards*, the coefficient is about ·5.

The paths of the particles at the vena contracta being nearly straight, there is little or no variation of pressure as we pass from the axis to the surface of the jet. We may therefore assume the velocity there to be uniform, and to have the value given by (14), where z now denotes the depth of the vena contracta below the surface of the liquid in the vessel. The rate of efflux is therefore

$$\sqrt{2gz} \cdot \rho S'.$$

32.   The calculation of the form of the issuing jet presents great difficulties, and has only been effected in one or two simple cases. (See Arts. 96, 97, below.) It is, however, easy to shew that the coefficient of contraction cannot (in the absence of friction) fall below the value $\frac{1}{2}$. For the pressure of the fluid at the walls of the vessel is approximately equal to the statical pressure $P + g\rho z$, except near the orifice, where on account of the velocity q becoming sensible, it is, by (13), somewhat less. Assuming it for the moment to be equal to the statical pressure, we see that the total horizontal pressure exerted on the fluid by the vessel is

$$PS + g\rho \iint z \, dS \dots\dots\dots\dots\dots\dots\dots(15),$$

where the integration extends over the area S of the orifice. The horizontal pressure exerted by any one element of the vessel's walls is in fact balanced by that due to an opposite element, except in the case of those elements which are opposite to the orifice. The first term of (15) is balanced by the pressure P of the atmosphere on the portion of fluid external to the vessel; so that the total horizontal force acting on the fluid is $g\rho \iint z \, dS$, or $g\rho \bar{z} S$, if $\bar{z}$ be the depth of the centre of inertia of the orifice. It is this force which produces the momentum with which the fluid leaves the vessel. The mass of fluid which in unit time passes the vena contracta is $\rho q S'$, and the momentum which this carries away with it is $\rho q^2 S'$. Hence, substituting the value of q from (14), we have

$$g\rho \bar{z} S = 2g\rho z S' \dots\dots\dots\dots\dots(16),$$

or, since z, $\bar{z}$ are nearly equal

$$S' : S = 1 : 2,$$

approximately. Since, however, the pressure on the wall is, near the orifice, sensibly less than the statical pressure $P + g\rho z$, the total horizontal force acting on the fluid somewhat exceeds the value (15). The left-hand side of (16) is therefore too small, and the ratio $S' : S$ is really greater than $\frac{1}{2}$.

The above theory is taken from a paper by Mr G. O. Hanlon, in the 3rd volume of the *Proceedings of the London Mathematical Society*, and from a note appended thereto by Professor Maxwell.

In one particular case, viz. where a short cylindrical tube, projecting inwards, is attached to the orifice, the assumption on which (16) was obtained is sensibly exact; and the value $\frac{1}{2}$ of the coefficient of contraction then agrees with experiment. Compare Art. 97.

33. *Example* 2. A gas flows through a small orifice from a receiver, in which the pressure is p_1 and the density ρ_1, into an open space where the pressure is p_2.[*]

We assume that the motion has become steady. In the receiver, at a distance from the orifice, we have $p = p_1$, $q = 0$, sensibly. This determines the value of C in equation (11). Neglecting the external forces, we find for the velocity of efflux

$$q = \sqrt{\left(\frac{2\gamma}{\gamma - 1}\right)\left(\frac{p_1}{\rho_1} - \frac{p_2}{\rho_2}\right)^{\frac{1}{2}}},$$

where ρ_2 is the density of the issuing gas at the vena contracta. If c be the velocity of sound in the gas of the receiver, we have (Chapter VIII.) $c^2 = \dfrac{\gamma p_1}{\rho_1}$; and therefore, taking account of (4), Art. 7,

$$q = \sqrt{\left(\frac{2}{\gamma - 1}\right)} \cdot c \left\{1 - \left(\frac{p_2}{p_1}\right)^{\frac{\gamma - 1}{\gamma}}\right\}^{\frac{1}{2}} \dots\dots\dots\dots(17).$$

---

[*] See Joule and Thomson, On the Thermal Effects of Fluids in Motion, *Proc. R. S.* May, 1856.

Also Rankine, *Applied Mechanics*, Arts. 637, 637 A.

The maximum velocity of efflux occurs when $p_2 = 0$, *i.e.* when the gas escapes into a vacuum; it is

$$\sqrt{\left(\frac{2}{\gamma - 1}\right)} \times \text{velocity of sound,}$$

or, for atmospheric air at 32° F., about 2413 feet per second.

The rate of escape of *mass* however depends on the value of $q\rho_2$, or

$$\sqrt{\left(\frac{2}{\gamma - 1}\right)} . c\rho_1 \left\{\left(\frac{p_2}{p_1}\right)^{\frac{2}{\gamma}} - \left(\frac{p_2}{p_1}\right)^{\frac{\gamma+1}{\gamma}}\right\}^{\frac{1}{2}} \quad \dots\dots\dots\dots(18),$$

which does not continually increase as p_2 diminishes, but attains a maximum when

$$\left(\frac{p_2}{p_1}\right)^{\frac{\gamma-1}{\gamma}} = \frac{2}{\gamma + 1}.$$

The velocity of efflux is then, by (17),

$$q = \sqrt{\left(\frac{2}{\gamma + 1}\right)} \times \text{velocity of sound,}$$

or, for atmospheric air at 32° F., about 997 feet per second.

The 'reduced velocity,' *i.e.* the velocity of a current of the density ρ_1 of the gas in the receiver which would convey matter at the same rate is got by dividing the expression (18) by ρ_1, and is, when a maximum, about 632 feet per second for air at 32° F.

34. *Example* 3. A mass of liquid rotates, under the action of gravity only, with constant angular velocity ω about the axis of z supposed drawn vertically upwards.

By hypothesis, $u = -\omega y, \quad v = \omega x, \quad w = 0$;

also $X = 0, \quad Y = 0, \quad Z = -g.$

The equation of continuity is identically satisfied, and the dynamical equations of motion become

$$-\omega^2 x = -\frac{1}{\rho}\frac{dp}{dx}, \quad -\omega^2 y = -\frac{1}{\rho}\frac{dp}{dy}, \quad 0 = -\frac{1}{\rho}\frac{dp}{dz} - g.$$

These have the common integral

$$\frac{p}{\rho} = \tfrac{1}{2}\omega^2 (x^2 + y^2) - gz + \text{const.}$$

The free surface, $p = \text{const.}$, is therefore a paraboloid of 'revolution about the axis of z, having its concavity upwards, and its latus rectum $= \dfrac{2g}{\omega^2}$.

Since $\dfrac{dv}{dx} - \dfrac{du}{dy} = 2\omega$, a velocity-potential does not exist. A motion of this kind could not be generated in a 'perfect' fluid, *i.e.* in one unable to sustain tangential stress. The fact that it can be realized with actual fluids shews that these are not 'perfect.'

35. *Example* 4. Instead of supposing the angular velocity ω to be uniform, let us suppose it to be a function of the distance r from that axis, and let us inquire what form must be assigned to this function in order that a velocity-potential may exist for the motion. We find

$$\frac{dv}{dx} - \frac{du}{dy} = 2\omega + r\frac{d\omega}{dr},$$

and that this may vanish we must have $\omega r^2 = \mu$, a constant. The velocity at any point is $= \dfrac{\mu}{r}$, so that the equation (9) becomes

$$\frac{p}{\rho} = \text{const.} - \tfrac{1}{2}\frac{\mu^2}{r^2},$$

if we suppose, for simplicity, that no external forces act. To find the velocity-potential ϕ, let us introduce polar co-ordinates r, θ. By Art. 27

$$\frac{d\phi}{dr} = \text{velocity along } r = 0,$$

$$\frac{d\phi}{rd\theta} = \text{velocity perpendicular to } r = \frac{\mu}{r},$$

so that

$$\phi = \mu\theta + \text{const.} = \mu \text{ arc tan} \frac{y}{x} + \text{const.} \quad \ldots\ldots\ldots (19).$$

We have here an instance of a *many-valued* or *cyclic* function. A function is said to be *single-valued* throughout any region of space when it is possible to assign to every point of that region a definite value of the function, in such a way that these values shall form a continuous series. This is not the case with the function in (19); for the value of ϕ there given, if it vary continuously, changes by $2\pi\mu$ as the point to which it refers describes a complete circuit round the origin, whereas a single-valued function would under the same circumstances return to its original value.

A function which like the above experiences a finite change of value when the point to which it refers describes a closed curve, returning to the point whence it started, is said to be many-valued or cyclic. The theory of many-valued velocity-potentials will be discussed in the next chapter.

36. *Example 5.* A mass of liquid filling a right circular cylinder moves from rest under the action of the forces

$$X = Ax + By, \quad Y = B'x + Cy, \quad Z = 0,$$

the axis of z being that of the cylinder.

Let us assume $u = -\omega y$, $v = \omega x$, $w = 0$, where ω is a function of t only. These values satisfy the equation of continuity and the boundary conditions. The dynamical equations become

$$\left.\begin{aligned}
-y\,\frac{d\omega}{dt} - \omega^2 x &= Ax + By - \frac{1}{\rho}\frac{dp}{dx}, \\
x\,\frac{d\omega}{dt} - \omega^2 y &= B'x + Cy - \frac{1}{\rho}\frac{dp}{dy}.
\end{aligned}\right\} \quad \ldots\ldots\ldots (20).$$

Differentiating the first of these with respect to y, and the second with respect to x and subtracting, we eliminate p, and find

$$\frac{d\omega}{dt} = \frac{B' - B}{2}.$$

The fluid therefore rotates as a whole about the axis of z with uniformly increasing angular velocity, except in the particular case when $B = B'$. To find p, we substitute the value of $\frac{d\omega}{dt}$ in (20) and integrate; thus we get

$$\frac{p}{\rho} = \tfrac{1}{2}\omega^2(x^2 + y^2) + \tfrac{1}{2}(Ax^2 + 2\beta xy + Cy^2) + \text{const.},$$

where $2\beta = B + B'$.

37. *Example 6.* Let $X = -\frac{\mu y}{r^2}$, $Y = \frac{\mu x}{r^2}$, $Z = 0$; the other circumstances remaining the same as in the preceding example.

Assuming $u = -\omega y$, $v = \omega x$, $w = 0$, where ω is a function of r $(= \sqrt{x^2 + y^2})$ and t only, we find

$$\left. \begin{array}{l} -y\dfrac{d\omega}{dt} - \omega^2 x = -\dfrac{\mu y}{r^2} - \dfrac{1}{\rho}\dfrac{dp}{dx}, \\[2mm] x\dfrac{d\omega}{dt} - \omega^2 y = \dfrac{\mu x}{r^2} - \dfrac{1}{\rho}\dfrac{dp}{dy}. \end{array} \right\} \quad \dots\dots\dots\dots(21).$$

Eliminating p, we obtain

$$2\frac{d\omega}{dt} + r\frac{d^2\omega}{dr\,dt} = 0.$$

The solution of this is

$$\omega = \frac{F(t)}{r^2} + f(r),$$

where F and f denote arbitrary functions. Since $\omega = 0$ when $t = 0$, we have

$$\frac{F(0)}{r^2} + f(r) = 0,$$

and therefore

$$\omega = \frac{F(t) - F(0)}{r^2} = \frac{\lambda}{r^2},$$

where λ is a function of t which vanishes for $t = 0$. Substituting in (21), and integrating, we find

$$\frac{p}{\rho} = \left(\mu - \frac{d\lambda}{dt}\right) \arctan \frac{y}{x} - \tfrac{1}{2}\omega^2 r^2 + \chi(t).$$

Since p is essentially a single-valued function, we must have $\dfrac{d\lambda}{dt} = \mu$, or $\lambda = \mu t$. Hence the fluid rotates with an angular velocity which varies inversely as the square of the distance from the axis, and increases uniformly with the time.

CHAPTER III.

38. THE present chapter is devoted mainly to an exposition of some general theorems relating to the class of motions already considered in Arts. 22—27; viz. those in which $u\,dx + v\,dy + w\,dz$ is an exact differential throughout a finite mass of fluid. It is convenient to begin with the following analysis, due to Stokes*, of the motion of a fluid element in the most general case.

The component velocities at the point (x, y, z) being u, v, w, those at an infinitely near point $(x + X, y + Y, z + Z)$ are

$$\left.\begin{aligned}
U &= u + \frac{du}{dx}\,X + \frac{du}{dy}\,Y + \frac{du}{dz}\,Z, \\
V &= v + \frac{dv}{dx}\,X + \frac{dv}{dy}\,Y + \frac{dv}{dz}\,Z, \\
W &= w + \frac{dw}{dx}\,X + \frac{dw}{dy}\,Y + \frac{dw}{dz}\,Z.
\end{aligned}\right\} \quad \dots\dots\dots\dots\dots(1).$$

If we write

$$a = \frac{du}{dx}, \quad b = \frac{dv}{dy}, \quad c = \frac{dw}{dz},$$

$$f = \tfrac{1}{2}\left(\frac{dw}{dy} + \frac{dv}{dz}\right), \quad g = \tfrac{1}{2}\left(\frac{du}{dz} + \frac{dw}{dx}\right), \quad h = \tfrac{1}{2}\left(\frac{dv}{dx} + \frac{du}{dy}\right),$$

$$\xi = \tfrac{1}{2}\left(\frac{dw}{dy} - \frac{dv}{dz}\right), \quad \eta = \tfrac{1}{2}\left(\frac{du}{dz} - \frac{dw}{dx}\right), \quad \zeta = \tfrac{1}{2}\left(\frac{dv}{dx} - \frac{du}{dy}\right),$$

equations (1) may be written

$$\left.\begin{aligned}
U &= u + aX + hY + gZ + \eta Z - \zeta Y, \\
V &= v + hX + bY + fZ + \zeta X - \xi Z, \\
W &= w + gX + fY + cZ + \xi Y - \eta X.
\end{aligned}\right\} \quad \dots\dots\dots\dots(2).$$

* *Camb. Phil. Trans.* Vol. VIII., 1845.

Hence the motion of a small element having the point (x, y, z) for its centre may be conceived as made up of three parts.

The first part, whose components are u, v, w, is a motion of *translation* of the element as a whole.

The second part, expressed by the second, third, and fourth terms on the right-hand side of the equations (2), is a motion such that every point on the quadric

$$aX^2 + bY^2 + cZ^2 + 2fYZ + 2gZX + 2hXY = \text{const.} \quad \ldots \ldots (3),$$

is moving in the direction of the normal to the surface. If we refer this quadric to its principal axes, the corresponding parts of the velocities parallel to these axes will be

$$U' = a'X', \quad V' = b'Y', \quad W' = c'Z'\ldots\ldots\ldots\ldots\ldots(4),$$

where $$a'X'^2 + b'Y'^2 + c'Z'^2 = \text{const.}$$

is what (3) becomes by the transformation. The formulæ (4) express that the length of every line in the element parallel to X' is being elongated at the rate (positive or negative) a', whilst lines parallel to Y' and Z' are being similarly elongated at the rates b' and c' respectively. Such a motion is called one of *pure strain* or *distortion*. The principal axes of (3) are called the axes of the strain or distortion.

The last two terms on the right-hand side of the equations (2) express a *rotation* of the element as a whole about an instantaneous axis; the component angular velocities of the rotation being ξ, η, ζ.

It can be shewn that the above resolution of the motion is unique. If we assume that the motion relative to the point (x, y, z) can be made up of a distortion and a rotation in which the axes and coefficients of the distortion and the axis and angular velocity of the rotation are arbitrary, then calculating the relative velocities $U - u$, $V - v$, $W - w$, we get expressions similar to those on the right-hand side of (2), but with arbitrary values of $a, b, c, f, g, h, \xi, \eta, \zeta$. Equating coefficients of X, Y, Z, however, we find that a, b, c, &c. must have the same values as before. Hence the directions of the axes of distortion, the rates of extension or contraction along them, and the axis and the angular velocity of rotation, at any point of the fluid, depend only on the state of relative motion at that point, and not on the position of the axes of reference.

When throughout a finite portion of a fluid mass we have ξ, η, ζ all zero, the motion of any element of that portion consists of a translation and a distortion only. We follow Thomson in calling the motion in such cases 'irrotational,' and that in all other cases 'rotational.'

39. The value of the integral $\int(udx + vdy + wdz)$, or, otherwise, $\int\left(u\dfrac{dx}{ds} + v\dfrac{dy}{ds} + w\dfrac{dz}{ds}\right)ds$, taken along any line $ABCD$, is called [*] the 'flow' of the fluid from A to D along that line. We shall denote it for shortness by $I(ABCD)$.

If A and D coincide, so that the line forms a closed curve, or circuit, the value of the integral is called the 'circulation' in that circuit. We denote it by $I(ABCA)$. If in either case the integration be taken in the opposite direction, the signs of $\dfrac{dx}{ds}$, &c., will be reversed, so that we have

$$I(AD) = -I(DA), \text{ and } I(ABCA) = -I(ACBA).$$

It is also plain that

$$I(ABCD) = I(AB) + I(BC) + I(CD).$$

Let us calculate the circulation in an infinitely small circuit surrounding the point (x, y, z). If $(x + X, y + Y, z + Z)$ be a point on the circuit, we have, by (2),

$$UdX + VdY + WdZ = d(UX + VY + WZ)$$
$$+ \tfrac{1}{2}d(aX^2 + bY^2 + cZ^2 + 2fYZ + 2gZX + 2hXY)$$
$$+ \xi(YdZ - ZdY) + \eta(ZdX - XdZ) + \zeta(XdY - YdX).$$

The first two lines of this expression, being exact differentials of single-valued functions, disappear when integrated round the cir-

---

[*] Thomson, On Vortex Motion. *Edin. Trans.* Vol. xxv., 1869.

cuit. Again $\int(YdZ - ZdY)$ is twice the area of the projection of the circuit on the plane yz, and therefore equal to $2ldS$, where dS is the area of the circuit, and l, m, n the direction-cosines of the normal to its plane. The coefficients of η and ζ give in the same way, on integration, $2mdS$ and $2ndS$, respectively. Hence, finally, the circulation round the circuit is

$$2\left(l\xi + m\eta + n\zeta\right)dS\dots\dots\dots\dots\dots\dots\dots(5),$$

or, twice the product of the area of the circuit into the component angular velocity of the fluid about the normal to its plane.

We have here tacitly made the convention that the direction of the normal to which l, m, n refer, and the direction in which the circulation in the circuit is estimated, are related in the same manner as the directions of advance and rotation in a right-handed screw *. ·

40. Any finite surface may be divided, by a double series of straight lines crossing it, into an infinite number of infinitely small elements. The sum of the circulations round the boundaries of these elements, taken all in the same sense, is equal to the circulation round the original boundary of the surface (supposed for the moment to consist of a single closed curve). For, in the sum in question, the flow along each side common to two elements

Fig. 2.

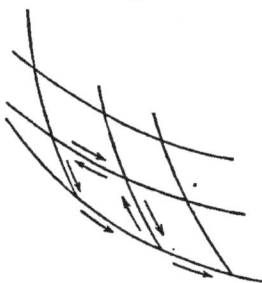

comes in twice, once for each element, but with opposite signs, and therefore disappears from the result. There remain then only the flows along those sides which are parts of the original boundary; whence the truth of the above statement.

* See Maxwell, *Electricity and Magnetism*, Art. 23.

Expressing this statement analytically we have, by (5),

$$\iint 2 \left(l\xi + m\eta + n\zeta \right) dS = \int (u\,dx + v\,dy + w\,dz)\ldots\ldots\ldots(6),$$

or, substituting the values of ξ, η, ζ from Art. 38,

$$\iint \left\{ l \left(\frac{dw}{dy} - \frac{dv}{dz} \right) + m \left(\frac{du}{dz} - \frac{dw}{dx} \right) + n \left(\frac{dv}{dx} - \frac{du}{dy} \right) \right\} dS$$

$$= \int (u\,dx + v\,dy + w\,dz)\ldots\ldots\ldots(7);$$

where the double-integral is taken over the surface, and the single-integral along the bounding curve. In these formulæ the quantities l, m, n are the direction-cosines of the normal drawn always on one side of the surface, which we may term the positive side; the direction of integration in the second member is then that in which a man walking on the surface, on the positive side of it, and close to the edge, must proceed so as to have the surface always on his left hand.

The theorem (6) or (7) may evidently be extended to a surface whose boundary consists of two or more closed curves, provided the integration in the second member be taken round each of these in the proper direction, according to the rule just given.

Fig. 3.

Thus, if the surface-integral in (6) extend over the shaded portion of the annexed figure, the directions in which the circulations in the several parts of the boundary are to be taken are shewn by the arrows, the positive side of the surface being that which faces the reader.

The value of the surface-integral taken over a *closed* surface is zero.

It should be noticed that (7) is a theorem of pure mathematics, and is true whatever functions u, v, w may be of x, y, z, provided only they be continuous over the surface*.

Irrotational Motion.

41. The rest of this chapter is devoted to the study of irrotational motion, as defined by the equations

$$\xi = 0, \quad \eta = 0, \quad \zeta = 0 \dots\dots\dots\dots\dots\dots\dots (8).$$

The existence and the properties of the velocity-potential in the various cases that may arise will appear as consequences of these equations.

Considering any region occupied by irrotationally-moving fluid, we see from (6) that the circulation is zero in every circuit which can be filled up by a continuous surface lying wholly in the region, or in other words capable of being contracted to a point without passing out of the region. Such a circuit is said to be 'evanescible.'

Again, let us consider two paths ACB, ADB, connecting two points A, B of the region, and such that either may by continuous variation be made to coincide with the other, without ever passing out of the region. Such paths are called 'mutually reconcileable.' Since the circuit $ACBDA$ is evanescible, we have

$$I(ACBDA) = 0, \text{ or since } I(BDA) = -I(ADB),$$

$$I(ACB) = I(ADB);$$

i.e. the flow is the same along any two reconcileable paths.

A region such that *all* paths joining any two points of it are mutually reconcileable is said to be 'simply-connected.' Such a region is that enclosed within a sphere, or that included between two concentric spheres. In what follows, as far as Art. 53, we contemplate only simply-connected regions.

42. The irrotational motion of a fluid within a simply-connected region is characterized by the existence of a single-valued

---

* It is not necessary that their differential coefficients should be continuous.

The theorem (7) is attributed by Maxwell to Stokes, *Smith's Prize Examination Papers for* 1854. The proof given above is due to Thomson, *l.c. ante*. For other proofs, see Thomson and Tait, *Natural Philosophy*, Art. 190 (j), and Maxwell, *Electricity and Magnetism*, Art. 24.

velocity-potential. Let ϕ denote the flow from some fixed point A to a variable point P, viz.

$$\phi = \int_A^P (udx + vdy + wdz) \dots\dots\dots\dots\dots (9).$$

The value of ϕ has been shewn to be independent of the path along which the integration is effected, provided it lie wholly within the region. Hence ϕ is a single-valued function of the position of P; let us suppose it expressed in terms of the co-ordinates (x, y, z) of that point. By displacing P through an infinitely short space parallel to each of the axes of co-ordinates in succession, we find

$$\frac{d\phi}{dx} = u, \quad \frac{d\phi}{dy} = v, \quad \frac{d\phi}{dz} = w,$$

i.e. ϕ is a velocity-potential, according to the definition of Art. 22.

The substitution of any other point B for A, as the lower limit in (9), simply adds an arbitrary constant to the value of ϕ, viz. the flow from B to A. The original definition of ϕ in Art. 22, and its physical interpretation in Art. 26, leave the function indeterminate to the extent of an additive constant.

As we follow the course of any stream-line the value of ϕ continually increases; hence in a simply-connected region the stream-lines cannot form closed curves.

43. The function ϕ with which we have here to do is, together with its first differential coefficients, by the nature of the case, finite, continuous, and single-valued at all points of the region considered. In the case of incompressible fluids, which we now proceed to consider more particularly, ϕ must also satisfy the equation of continuity, (5) of Art. 25, or as we shall write it, for shortness,

$$\nabla^2 \phi = 0 \dots\dots\dots\dots\dots\dots\dots (10),$$

at every point of the region. Hence ϕ is now subject to mathematical conditions identical with those satisfied by the potential of masses attracting or repelling according to the law of the inverse square of the distance, at all points external to such masses; so that many of the results proved in the theories of Attractions, Statical Electricity, &c., have also a hydrodynamical application.

We proceed to develope those which are most important from this point of view.

44. The proof of (10) given in Art. 12 is based essentially on the consideration that since the fluid is incompressible the total volume which enters any element $dx\,dy\,dz$ in unit time is zero. To apply the same principle to a finite region occupied entirely by liquid, let dS be an element of the surface of the region, dn an element of the normal to it drawn inwards. By Art. 27 $\frac{d\phi}{dn}$ is the inward velocity of the fluid normal to the surface, and therefore $\frac{d\phi}{dn}\,dS$ is the volume which in unit time enters the region across the element dS. Hence

$$\iint \frac{d\phi}{dn}\,dS = 0 \quad \dotfill (11),$$

the integration extending over the whole boundary of the region. Equations (10) and (11), expressing the same fact, must be mathematically equivalent; see Art. 64.

The stream-lines drawn through the various points of an infinitesimal circuit constitute a tube, which may be called a 'tube of flow.' The product of the velocity (q) into the cross-section (σ) is the same at all points of such a tube.

We may, if we choose, regard the whole space occupied by the fluid as made up of tubes of flow, and suppose the size of the tubes so chosen that the product $q\sigma$ is the same for each. The value of the integral $\iint \frac{d\phi}{dn}\,dS$ taken over any surface is then proportional to the number of tubes which cross that surface. If the surface be closed, the equation (11) expresses the fact that as many tubes cross the surface inwards as outwards. Hence a stream-line cannot begin or end at a point of the fluid.

45. The function ϕ cannot be a maximum or minimum at a point in the interior of the fluid; for, if it were, we should have $\frac{d\phi}{dn}$ everywhere positive, or everywhere negative, over a small closed surface surrounding the point in question. Each of these suppositions is inconsistent with (11).

Further, the velocity cannot be a *maximum* at a point in the interior of the fluid. For let the axis of x be taken parallel to the direction of the velocity at any point P. The equation (10), and therefore also the equation (11), is satisfied if we write $\frac{d\phi}{dx}$ for ϕ. The above argument then shews that $\frac{d\phi}{dx}$ cannot be a maximum at P. Hence there must be some point in the immediate neighbourhood of P for which $\frac{d\phi}{dx}$ has a greater value, and therefore *a fortiori*, for which $\left\{\left(\frac{d\phi}{dx}\right)^2 + \left(\frac{d\phi}{dy}\right)^2 + \left(\frac{d\phi}{dz}\right)^2\right\}^{\frac{1}{2}}$ is greater than $\frac{d\phi}{dx}$, *i.e.* the velocity of the fluid at some neighbouring point is greater than that at P*.

On the other hand, the velocity may be a *minimum* at some point of the fluid. In fact, taking any case of fluid motion, let us impress on the whole mass a velocity equal and opposite to that at any point P of it. In the resulting motion the velocity at P will be zero, and therefore a minimum.

46. Let us apply (11) to the boundary of a finite spherical portion of the liquid. If r denote the distance of any point from the centre of the sphere, $d\varpi$ the elementary solid angle subtended at the centre by an element dS of the surface, we have

$$\frac{d\phi}{dn} = -\frac{d\phi}{dr},$$

and $dS = r^2 d\varpi$. Omitting the constant factor r^2, (11) becomes

$$\iint \frac{d\phi}{dr} d\varpi = 0,$$

or

$$\frac{d}{dr} \iint \phi\, d\varpi = 0 \dots\dots\dots\dots\dots\dots\dots(12).$$

Since $\frac{1}{4\pi} \iint \phi\, d\varpi$, or $\frac{1}{4\pi r^2} \iint \phi\, dS$, is the mean value of ϕ over the surface of the sphere, (12) shews that this mean value is inde-

---

* This theorem was set by Prof. Maxwell as a question in the Mathematical Tripos, 1873. The above proof is taken from Kirchhoff, *Vorlesungen über Mathematische Physik. Mechanik*, p. 186. Another proof is given below, Art. 64.

pendent of the radius. It is therefore the same for any sphere, concentric with the former one, which can be made to coincide with it by gradual variation of the radius, without ever passing out of the region occupied by the irrotationally moving liquid. We may therefore suppose the sphere contracted to a point, and so obtain a simple proof of the theorem, first given by Gauss in his memoir* on the theory of Attractions, that the mean value of ϕ over any spherical surface throughout the interior of which (10) is satisfied, is equal to its value at the centre.

The theorem, proved in Art. 45, that ϕ cannot be a maximum or a minimum at a point in the interior of the fluid, is an obvious consequence of the above.

Again, let us suppose that the region occupied by the irrotationally moving fluid is 'periphractic,'† i.e. that it is limited internally by one or more closed surfaces, and let us apply (11) to the space included between one (or more) of these internal boundaries, and a spherical surface completely enclosing it and lying wholly in the fluid. If $4\pi M$ denote the total flux inwards across the internal boundary of this space, we find, with the notation as before,

$$\iint \frac{d\phi}{dr}\, dS = 4\pi M,$$

the surface integral extending over the sphere only. This may be written

$$\frac{1}{4\pi} \frac{d}{dr} \iint \phi\, d\varpi = \frac{M}{r^2},$$

whence

$$\frac{1}{4\pi r^2} \iint \phi\, dS = \frac{1}{4\pi} \iint \phi\, d\varpi = -\frac{M}{r} + C \ldots\ldots\ldots\ldots(13).$$

That is, the mean value of ϕ over any spherical surface drawn under the above-mentioned conditions is equal to $-\dfrac{M}{r} + C$, where r is the radius, M an absolute constant, and C a quantity which is independent of the radius but may vary with the position of the centre‡.

---

* *Werke*, t. 5, p. 199. A translation of the memoir is given in Taylor's *Scientific Memoirs*, Vol. III.

† See Maxwell, *Electricity and Magnetism*, Arts. 18, 22.

‡ It is understood, of course, that the spherical surfaces to which this statement

If however the original region throughout which the irrotational motion holds be unlimited externally, and if the first (and therefore all the higher) derivatives of ϕ vanish at infinity, then C is the same for *all* spherical surfaces enclosing the whole of the internal boundaries. For if such a sphere be displaced parallel to x^*, without alteration of size, the rate at which C varies in consequence of this displacement is, by (13), equal to the mean value of $\dfrac{d\phi}{dx}$ over the surface. Since $\dfrac{d\phi}{dx}$ vanishes at infinity, we can by taking the sphere large enough make the latter mean value as small as we please. Hence C is not altered by a displacement of the centre of the sphere parallel to x. In the same way we see that C is not altered by a displacement parallel to y or z; *i.e.* it is absolutely constant.

If the internal boundaries be such that the total flux across them is zero, *e.g.* if they be the surfaces of solids, or of portions of incompressible fluid whose motion is rotational, we have $M = 0$, so that the mean value of ϕ over *any* spherical surface enclosing them all is the same.

47. (α) If ϕ be constant over the boundary of any simply-connected region occupied by liquid moving irrotationally, it has the same constant value throughout the interior of that region. For if not constant it would necessarily have a maximum or a minimum value at some point of the region.

Otherwise: we have seen in Arts. 42, 44 that the stream-lines cannot begin or end at any point of the region, and that they cannot form closed curves lying wholly within it. They must therefore traverse the region, beginning and ending on its boundary. In our case this is however impossible, for a stream-line always proceeds from places where ϕ is less to places where it is greater, whereas ϕ is, by hypothesis, constant over the boundary. Hence there can be no motion, *i.e.*

$$\frac{d\phi}{dx} = 0, \quad \frac{d\phi}{dy} = 0, \quad \frac{d\phi}{dz} = 0,$$

and therefore ϕ is constant and equal to its value at the boundary.

applies are 'reconcileable' (in a sense analogous to that of Art. 41) with one another.

* This step is taken from Kirchhoff, *Vorlesungen über Math. Physik. Mechanik,* p. 191.

(β) Again, if $\frac{d\phi}{dn}$ be zero at every point of the boundary of such a region as is above described, ϕ will be constant throughout the interior. For the condition $\frac{d\phi}{dn} = 0$, expresses that no stream-lines enter or leave the region, but that they are all contained within it. This is however, as we have seen, inconsistent with the other conditions which the stream-lines must conform to. Hence, as before, there can be no motion, and ϕ is constant.

This theorem may be otherwise stated as follows: no irrotational motion of a liquid can take place throughout a simply-connected region bounded entirely by fixed rigid walls.

(γ) Again, let the boundary of the region considered consist partly of surfaces S over which ϕ has a given constant value, and partly of other surfaces Σ over which $\frac{d\phi}{dn} = 0$. By the previous argument, no stream-lines can pass from one point to another of S, and none can cross Σ. Hence no stream-lines exist; ϕ is therefore constant as before, and equal to its value at S.

48. Recalling the dynamical interpretation of ϕ given in Art. 26, we see that the first theorem of Art. 47 asserts that a uniform impulsive pressure applied to the boundary of a liquid mass produces no motion. This is otherwise obvious.

A non-uniform impulsive pressure applied to the boundary of the mass will of course generate some definite motion. Further, it appears highly probable that this motion will be everywhere finite and continuous throughout the mass; although if this be the case right up to the boundary, it is necessary that the given surface-value of the impulsive pressure should be continuous, and should also have its rate of variation from point to point of the boundary everywhere finite and continuous. We are thus led to the following analytical theorem:

(α) There exists a single-valued function ϕ which satisfies the equation $\nabla^2\phi = 0$ at every point of a finite region of space, which is with its first derivatives finite and continuous throughout that region, and which has a given arbitrary value at every point of the boundary. If the finiteness and continuity of ϕ hold up to the boundary, the arbitrarily given surface-value must, with its rate of variation from point to point, be finite and continuous.

Again, let us consider a mass of liquid initially at rest and enveloped by a perfectly smooth flexible membrane which it just fills; and let us suppose that every point of the membrane is suddenly moved with a given normal velocity $\frac{d\phi}{dn}$, which must of course satisfy the condition $\iint \frac{d\phi}{dn}\, dS = 0$, where the integration extends over the whole membrane. Since some definite motion of the mass must ensue, we are led, on the same kind of evidence as before, to the following analytical theorem;

(β)   There exists a single-valued function ϕ which satisfies the equation $\nabla^2\phi = 0$ at every point of a given region, which is with its first derivatives finite and continuous throughout that region, and which has at every point of the boundary its rate of variation $\frac{d\phi}{dn}$ in the direction of the normal equal to a given arbitrary value, subject to the condition $\iint \frac{d\phi}{dn}\, dS = 0$. If the finiteness and continuity of ϕ and its derivatives hold right up to the boundary, the given surface-values of $\frac{d\phi}{dn}$ must be continuous.   See Art. 84.

(γ)   Lastly, combining the two modes of genesis of motion described above, we are led to enunciate the theorem that a single-valued function ϕ exists which satisfies $\nabla^2\phi = 0$ at all points of a given region, which is with its first derivatives finite and continuous throughout the region, and which has a given arbitrary value over one part of the boundary and gives an assigned arbitrary value of $\frac{d\phi}{dn}$ over the rest of the boundary.

49.   The physical considerations adduced in support of the above theorems are due to Thomson*. The theorem (α) was originally stated by Green†, and based by him on electrical considerations. An analytical proof, couched however in the language of the

---

* *Reprint of Papers on Electrostatics and Magnetism*, Article xxviii.

† *An essay on Electricity and Magnetism* (1828), § 5.

theory of Attractions, was first given by Gauss (*l. c.* Art. 46). Another method of proof, purely analytical in form, and applicable to theorems (β) and (γ) as well, was given by Thomson[*] in 1848. As this method is not free from difficulty, it is not reproduced here. It is however perfectly easy to shew analytically that in theorems (α) and (γ) of Art. 48 ϕ is completely determinate, *i.e.* that there is only one function satisfying the conditions stated, and that in (β) ϕ is determinate save as to an additive constant.

In the first place, if ϕ_1, ϕ_2, ϕ_3, &c. be velocity-potentials of possible states of motion throughout any given region, then

$$\phi = A_1\phi_1 + A_2\phi_2 + A_3\phi_3 + \&c.,$$

where A_1, A_2, A_3, &c. are any constants, is the velocity-potential of a possible state of motion (throughout the region). This follows from the linearity of (10).

Now, if possible, let there be two single-valued functions ϕ_1, ϕ_2, each satisfying the conditions of (α) with respect to any region. Then $\phi_1 - \phi_2$ satisfies (10) throughout this region and is zero at every point of the boundary. It is therefore, by Art. 47, zero throughout; *i.e.*, ϕ_1, ϕ_2 are identical.

Again, if it be possible, let ϕ_1, ϕ_2 be two single-valued functions each satisfying the conditions of (β) with respect to any region. Then $\phi_1 - \phi_2$ satisfies (10) throughout this region, and makes

$$\frac{d}{dn}(\phi_1 - \phi_2) = 0$$

all over the boundary. Hence by Art. 47 we have $\phi_1 - \phi_2$ constant throughout the region; and therefore the motion, which is determined by the derivatives of ϕ, is the same in each case.

Lastly, if ϕ_1, ϕ_2 be two single-valued functions satisfying the conditions of (γ), it is seen in the same way that we must have $\phi_1 - \phi_2 = 0$.

50.   A class of cases of great importance, but not strictly included in the scope of the foregoing theorems, are those where the region occupied by the liquid extends to infinity, but is bounded internally by one or more closed surfaces. We assume, for the

[*] See *Reprint*, Article XIII. Also Thomson and Tait, *Natural Philosophy*, Appendix A (*d*). This method is often attributed by German writers to Dirichlet.

present, that this region is simply-connected, and that ϕ is therefore single-valued.

If ϕ be constant over the internal boundary of the region, and tend everywhere to the same constant value at an infinite distance from the internal boundary, it is constant throughout the region. For otherwise ϕ would be a maximum or a minimum at some point of the region.

We infer, exactly as in Art. 49, that if ϕ be given arbitrarily over the internal boundary, and have a given constant value at infinity, its value is everywhere determinate.

Of more importance in our present subject is the theorem that if the normal velocity $\dfrac{d\phi}{dn}$ be zero at every point of the internal boundary, and if the fluid be at rest at infinity, then ϕ is everywhere constant. We cannot however infer this at once from the proof of the corresponding theorem in Art. 47. It is true that we may suppose the region limited externally by an infinitely large surface at every point of which $\dfrac{d\phi}{dn}$ is infinitely small; but it is conceivable that the integral $\iint \dfrac{d\phi}{dn} dS$ taken over a portion of this surface might be finite, in which case the investigation referred to would fail. We proceed, therefore, somewhat indirectly, as follows.

51. Since the velocity tends to the limit zero at an infinite distance from the internal boundary (S, say), it must be possible to draw a closed surface Σ, completely enclosing S, beyond which the velocity is everywhere less than a certain small value ϵ, which value may, by making Σ large enough, be made as small as we please. Now in any direction from S let us take a point P at such a distance beyond Σ that the solid angle which Σ subtends at it is infinitely small; and with P as centre describe two spheres, one just excluding, the other just including S. We shall prove that the mean value of ϕ over each of these spheres is, within an infinitely small amount, the same. For if Q, Q' be points of these spheres on a common radius PQQ', then if Q, Q' fall within Σ the corresponding values of ϕ may differ by a finite amount; but since the portion of either spherical surface which falls within Σ is an infinitely small fraction of the whole, no finite difference

in the mean values can arise from this cause. On the other hand, when Q, Q' fall without Σ, the corresponding values of ϕ cannot differ by so much as $\epsilon . QQ'$, for ϵ is by definition a superior limit to the rate of variation of ϕ. Hence, the mean values of ϕ over the two spherical surfaces must differ by less than $\epsilon . QQ'$. Since QQ' is finite, whilst ϵ may by taking Σ large enough be made as small as we please, the difference of the mean values may, by taking P sufficiently distant, be made infinitely small.

Now we have seen in Art. 46 that the mean value of ϕ over the inner sphere is equal to its value at P, and that the mean value over the outer sphere is (since $M=0$) equal to a constant quantity C. Hence, ultimately, the value of ϕ at infinity tends everywhere to the constant value C.

The same result holds even if the normal velocity $\dfrac{d\phi}{dn}$ be not zero over the internal boundary; for in the theorem of Art. 46 M is divided by r, which is in our case infinite.

52. The theorem stated at the end of Art. 50 is now obvious. For, under the conditions there stated, no stream-lines can begin or end on the internal boundary. Hence, any stream-lines which exist must come from an infinite distance, traverse the region, and pass off again to infinity; *i.e.* they perform infinitely long courses between places where ϕ has, within an infinitely small amount, the same value C, which is impossible. Hence no stream-lines exist, or in other words there is no motion.

We derive, exactly as in Art. 49, the important theorem that if $\dfrac{d\phi}{dn}$ be given at every point of the internal boundary, and if the velocity be zero at infinity, the motion is everywhere determinate.

53. Before discussing the properties of irrotational motion in multiply-connected regions we must examine more in detail the nature and classification of such regions. In the following synopsis of this branch of the geometry of position we recapitulate for the sake of completeness one or two definitions already given.

On Multiply-Connected Regions.

We consider any connected region of space, enclosed by boundaries. A region is 'connected' when it is possible to pass from any one point of it to any other by an infinity of paths, each of which lies wholly in the region.

Any two such paths, or any two circuits, which can by continuous variation be made to coincide without ever passing out of the region, are said to be 'mutually reconcileable.' Any circuit which can be contracted to a point without passing out of the region is said to be 'evanescible.' Two reconcileable paths, combined, form an evanescible circuit. If two paths or two circuits be reconcileable, it must be possible to connect them by a continuous surface, which lies wholly within the region, and of which they form the complete boundary; and conversely.

It is further convenient to distinguish between 'simple' and 'multiple' non-evanescible circuits. A 'multiple' circuit is one which can by continuous variation be made to appear, in whole or in part, as the repetition of another circuit a certain number of times. A 'simple' circuit is one with which this is not possible. There is no distinction between simple and multiple evanescible circuits.

A 'barrier,' or 'diaphragm,' is a surface drawn across the region, and limited by the line or lines in which it meets the boundary. Hence a barrier is necessarily a connected surface, and cannot consist of two or more detached portions.

A 'simply-connected' region is one such that all paths joining any two points of it are reconcileable, or such that all circuits drawn within it are evanescible.

A 'doubly-connected' region is one such that two irreconcileable paths, and no more, can be drawn between any two points A, B of it; viz. any other path joining AB is reconcileable with one of these, or with a combination of the two taken each a certain number of times. In other words, the region is such that one (simple) non-evanescible circuit can be drawn in it, whilst all other circuits are either reconcileable with this (repeated, if necessary), or are evanescible. As an example of a doubly-connected region we may take that enclosed by an anchor-ring, or that external to such a ring and extending to infinity.

Generally, a region such that n irreconcileable paths, and no more, can be drawn between any two points of it, or such that $n - 1$

(simple) irreconcileable non-evanescible circuits, and no more, can be drawn in it, is said to be 'n-ply-connected.'

The shaded portion of Fig. 3, Art. 40, is a triply-connected space of two dimensions.

It is shewn in note (B) that the above definition of an n-ply-connected space is self-consistent. In such simple cases as $n = 2$, $n = 3$, this is sufficiently obvious without demonstration.

54. Let us suppose, now, that we have an n-ply-connected region, with $n - 1$ simple independent non-evanescible circuits drawn in it. It is possible to draw a barrier meeting any one of these circuits in one point only, and not meeting any of the $n - 2$ remaining circuits*. A barrier drawn in this manner does not destroy the continuity of the region, for the interrupted circuit remains as a path leading round from one side of the barrier to the other. The order of connection of the region is however reduced by unity; for every circuit drawn in the modified region must be reconcileable with one or more of the $n - 2$ circuits not met by the barrier.

A second barrier, drawn in the same manner, will reduce the order of connection again by one, and so on; so that by drawing $n - 1$ barriers we can reduce the region to a simply-connected one.

A simply-connected region is divided by a barrier into two separate parts; for otherwise it would be possible to pass from a point on one side the barrier to an adjacent point on the other side by a path lying wholly within the region, which path would in the original region form a non-evanescible circuit.

Hence in an n-ply-connected region it is possible to draw $n - 1$ barriers, and no more, without destroying the continuity of the region. We might, if we had so chosen, have taken this property as the definition of an n-ply-connected space. We leave it as an exercise for the student to prove that this definition is free from ambiguity, and that it is equivalent to the former one.

Irrotational Motion in Multiply-connected Spaces.

55. The circulation is the same in any two reconcileable circuits $ABCA$, $A'B'C'A'$ drawn in a region occupied by fluid moving

---

* In simple cases this is obvious. For a general proof see note (B).

irrotationally. For the two circuits may be connected by a continuous surface lying wholly within the region; and if we apply the theorem of Art. 40 to this surface, we have, remembering the rule as to the direction of integration round the boundary,

$$I\,(ABCA) + I\,(A'C'B'A') = 0,$$

or
$$I\,(ABCA) = I\,(A'B'C'A').$$

If a circuit $ABCA$ be reconcileable with two or more circuits $A'B'C'A'$, $A''B''C''A''$, &c., combined, we can connect all these circuits by a continuous surface which lies wholly within the region, and of which they form the complete boundary. Hence

$$I(ABCA) + I\,(A'C'B'A') + I(A''C''B''A'') + \&c. = 0,$$

or
$$I\,(ABCA) = I\,(A'B'C'A') + I\,(A''B''C''A'') + \&c.;$$

i.e. the circulation in any circuit is equal to the sum of the circulations in the several members of any set of circuits with which it is reconcileable.

Let the order of connection of the region be $n + 1$, so that n independent simple non-evanescible circuits $a_1, a_2, \ldots a_n$ can be drawn in it; and let the circulations in these be $\kappa_1, \kappa_2, \ldots \kappa_n$, respectively. The sign of any κ will of course depend on the direction of integration round the corresponding circuit; let the direction in which κ is estimated be called the positive direction in the circuit. The value of the circulation in any other circuit can now be found at once. For the given circuit is necessarily reconcileable with some combination of the circuits $a_1, a_2, \ldots a_n$; say with a_1 taken p_1 times, a_2 taken p_2 times and so on, where of course any p is negative when the corresponding circuit is taken in the negative direction. The required circulation then is

$$p_1\kappa_1 + p_2\kappa_2 + \ldots + p_n\kappa_n \ldots\ldots\ldots\ldots\ldots\ldots(14).$$

Since any two paths joining two points A, B of the region together form a circuit, it follows that the values of the flow in the two paths differ by a quantity of the form (14), where, of course, in particular cases some or all of the p's may be zero.

56.　Let ϕ be the flow from a fixed point A to a variable point P, viz.

$$\phi = \int_A^P (u\,dx + v\,dy + w\,dz) \dots\dots\dots\dots(15).$$

So long as the path of integration from A to P is not specified, ϕ is indeterminate to the extent of a quantity of the form (14).

If however n barriers be drawn in the manner explained in Art. 54, so as to reduce the region to a simply-connected one, and if the path of integration in (15) be restricted to lie within the region as thus modified (*i.e.* it is not to cross any of the barriers), then ϕ becomes a single-valued function, as, in Art. 42. It is continuous throughout the modified region, but its values at two adjacent points on opposite sides of a barrier differ by $\pm\,\kappa$. To derive the value of ϕ when the integration is taken along any path in the unmodified region we must add the quantity (14), where any p denotes the number of times this path crosses the corresponding barrier. A crossing in the positive direction of the circuits interrupted by the barrier is here counted as positive, a crossing in the opposite direction as negative.

By displacing P through an infinitely short space parallel to each of the co-ordinate axes in succession, we find

$$u = \frac{d\phi}{dx}, \quad v = \frac{d\phi}{dy}, \quad w = \frac{d\phi}{dz};$$

so that ϕ satisfies the definition of a velocity-potential, Art. 22. It is now however a many-valued or cyclic function; *i.e.* it is not .possible to assign to every point of the original region a definite value of ϕ, such values forming a continuous system. On the contrary, whenever P describes in the region a non-evanescible circuit, ϕ will not, in general, return to its original value, but will differ from it by a quantity of the form (14). The quantities $\kappa_1, \kappa_2, \dots \kappa_n$ are called by Thomson the 'cyclic constants' of ϕ.

57.　The foregoing theory is illustrated by Ex. 4, Art. 35. The formulæ there given make the velocity infinite at points on the axis of z, which must therefore be excluded from the region to which our theorems apply. This region becomes thereby doubly-connected, for we can connect any two points A, B of it by two

irreconcileable paths passing on opposite sides of the axis, *e.g.*
ACB, ADB in the figure.   The portion of the plane *zx* for which

Fig. 4.

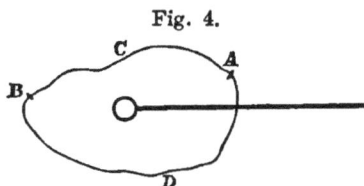

x is positive may be taken as a barrier, and the region is thus
made simply-connected.   The circulation in any circuit meeting
this barrier once only, *e.g.* in *ACBDA*, is $\int_0^{2\pi} \frac{\mu}{r} . r d\theta$, or $2\pi\mu$.   That
in any circuit not meeting the barrier is zero.   In the modified
region ϕ may be put equal to a single-valued function, viz. $\mu\theta$,
but its value on the positive side of the barrier is zero, that at an
adjacent point on the negative side is $2\pi\mu$.

More complex illustrations of irrotational motion in multiply-
nected spaces will present themselves in the next chapter.

58.   Before proceeding further we may briefly indicate a somewhat different method of presenting the above theory.

Starting from the existence of a velocity-potential as the
characteristic of the class of motions which we wish to study, and
adopting the second definition of an $n + 1$-ply-connected region,
given in Art. 54, we remark that in a simply-connected region
every equipotential surface must either be a closed surface, or
else form a barrier dividing the region into two separate parts.
Hence, supposing the whole system of such surfaces drawn, we see
that if a closed curve cross any given equipotential surface once it
must cross it again, and in the opposite direction.   Hence, corresponding to any element of the curve, included between two
consecutive equipotential surfaces, we have a second element such
that the flow along it, being equal to the difference between the
corresponding values of ϕ, is equal and opposite to that along the
former; so that the circulation in the whole circuit is zero.

If however the region be multiply-connected, an equipotential
surface may form a barrier without dividing it into two separate
parts.   Let as many such surfaces be drawn as it is possible to draw

without destroying the continuity of the region. The number of these cannot, by definition, be greater than n. Every other equipotential surface which is not closed will be reconcileable (in an obvious sense) with one or more of these barriers. A curve drawn from one side of one of these barriers round to the other, without meeting any of the remaining barriers, will cross every surface reconcileable with it an odd number of times, and every other surface an even number of times. Hence the circulation in the circuit thus formed will not vanish, and ϕ will be a cyclic function.

In the method adopted above we have based the whole theory on the equations

$$\frac{dw}{dy} - \frac{dv}{dz} = 0, \quad \frac{du}{dz} - \frac{dw}{dx} = 0, \quad \frac{dv}{dx} - \frac{du}{dy} = 0\ldots\ldots\ldots(16),$$

and have deduced the existence and properties of the velocity-potential in the various cases as necessary consequences of these. In fact, Arts. 41, 42, and 53—56, may be regarded as a treatise on the integration of this system of differential equations.

The integration of (16), when we have, on the right-hand side, instead of zero known functions of x, y, z, will be treated in Chapter VI.

59. If the density of the fluid be either constant or a function of the pressure only, and if the external impressed forces have a single-valued potential, the cyclic constants of ϕ do not alter with the time. For if ϕ_0 be the initial value of the velocity-potential, we have, Art. 23,

$$\phi = \phi_0 + \chi,$$

where, Art. 19 (25),

$$\chi = \int_0^t \left[-V - \int \frac{dp}{\rho} + \tfrac{1}{2} \left\{ \left(\frac{dx}{dt}\right)^2 + \left(\frac{dy}{dt}\right)^2 + \left(\frac{dz}{dt}\right)^2 \right\} \right] dt.$$

Under the circumstances stated χ is a single-valued function, and the cyclic constants of ϕ are the same as those of ϕ_0. In other words the circulations in the several circuits of the region occupied by the fluid are constant.

This is otherwise evident from Art. 25 (3), which shews that $\frac{d\phi}{dt}$ is single-valued, and that therefore the cyclic constants of ϕ cannot alter.

In Examples 5 and 6, Arts. 36, 37, we had instances to which the above result is not applicable; the reason being that in Ex. 5 the external forces have not a potential, whilst in Ex. 6 their potential is itself a cyclic function.

60. Proceeding now, as in Art. 43, to the particular case of an incompressible fluid, we remark that whether ϕ be many-valued or not, its first derivatives $\frac{d\phi}{dx}$, $\frac{d\phi}{dy}$, $\frac{d\phi}{dz}$, and therefore all the higher derivatives, are essentially single-valued functions, so that ϕ will still satisfy the equation of continuity

$$\nabla^2\phi = 0 \dots\dots\dots\dots\dots\dots\dots\dots(10),$$

and the equivalent form

$$\iint \frac{d\phi}{dn}\, dS = 0 \dots\dots\dots\dots\dots\dots(11),$$

where the surface-integration extends over the whole boundary of any portion of the fluid.

In the theorems of Arts. 45 and 46 the spaces to which (11) is applied are simply-connected, so that it is allowable to suppose ϕ single-valued throughout them even when the region of which they form a ·part is multiply-connected. On this understanding the theorems in question still hold when ϕ is a cyclic function.

The theorem (α) of Art. 47, viz. that ϕ must be constant throughout the interior of any region at every point of which (10) is satisfied if it be constant over the boundary, still holds when the region is multiply-connected. For ϕ, being constant over the boundary, is necessarily single-valued.

The remaining theorems of Art. 47, being based on the assumption that the stream-lines cannot form closed curves, are however no longer exact. We must introduce the additional condition that the circulation is to be zero in each circuit of the region.

The theorems of Art. 48 also call for modification.   The proper extension of (β) is as follows :

61.   A function ϕ exists which satisfies (10) throughout a given $n + 1$-ply-connected region, which has any given cyclic constants $\kappa_1, \kappa_2, \ldots \kappa_n$ corresponding to the n independent non-evanescible circuits capable of being drawn in the region, and which is such that its rate of variation $\dfrac{d\phi}{dn}$ in the direction of the normal has a given value at every point of the boundary.   These arbitrary values of $\dfrac{d\phi}{dn}$ must of course fulfil the condition $\iint\dfrac{d\phi}{dn}\,dS = 0$.

We follow Thomson in marshalling the following physical considerations in support of this theorem.   Let us suppose the region occupied by incompressible fluid of unit density enclosed in a perfectly smooth and flexible membrane.   Further, let n barriers be drawn, as in Art. 54, so as to reduce the region to a simply-connected one, and let their places be occupied by similar membranes, infinitely thin, and destitute of inertia.   The fluid being initially at rest, let each element of the first-mentioned membrane be suddenly moved inwards with the given (positive or negative) normal velocity $\dfrac{d\phi}{dn}$, whilst uniform impulsive pressures $\kappa_1, \kappa_2, \ldots \kappa_n$ are applied to the positive sides of the respective barrier-membranes.   Some definite motion of the fluid will ensue, characterized by the following properties :

(a)   It is irrotational, being generated from rest ;

(b)   The normal velocity at every point of the original boundary has the assigned value ;

(c)   The values of the impulsive pressure, and therefore of the velocity-potential, at two adjacent points on opposite sides of a barrier-membrane, differ by the corresponding value of κ, which is constant over the barrier ;

(d)   The motion on one side of a barrier is continuous with that on the other.

To prove the last statement we remark, first, that the velocities normal to the barrier at two adjacent points on opposite sides of it are the same, being each equal to the normal velocity of the adjacent portion of the membrane. Again, if P, Q be two consecutive points on a barrier, and if the corresponding values of ϕ be on one side ϕ_P, ϕ_Q, and on the other ϕ'_P, ϕ'_Q, we have, by (c)

$$\phi_P - \phi'_P = \kappa = \phi_Q - \phi'_Q,$$

or

$$\phi_Q - \phi_P = \phi'_Q - \phi'_P,$$

i.e., if $PQ = ds$,

$$\frac{d\phi}{ds} = \frac{d\phi'}{ds}.$$

Hence the tangential velocities at two adjacent points on opposite sides of a barrier also agree. If then we suppose the barrier-membranes to be liquefied immediately after the impulse, we obtain a state of irrotational motion satisfying the conditions stated at the head of this article*.

62. It is easy to shew analytically that the said conditions completely determine ϕ, save as to an additive constant. For, if possible, let there be two functions ϕ_1, ϕ_2 each satisfying the conditions. Since ϕ_1, ϕ_2 have the same cyclic constants, $\phi = \phi_1 - \phi_2$ is a single-valued function, which moreover satisfies (10) throughout the region, and makes $\frac{d\phi}{dn} = 0$ at every point of the boundary. Hence Art. 47 (β) applies, and shews that ϕ is constant.

Hence the irrotational motion throughout an $n + 1$-ply-connected space is determinate when we know the value of the normal velocity at every point of the boundary, and also the value of the circulation in each of the n independent circuits which can be drawn in that space.

The following theorem, which now replaces that of Art. 52, is proved in like manner.

The irrotational motion through an $n + 1$-ply-connected region extending to infinity, but limited internally by one or more closed surfaces, is made fully determinate by the following conditions:

---

* The modifications necessary in theorems (a) and (γ) of Art. 48 are passed over, as of little interest in our present subject.

(*a*) The normal velocity has a prescribed value at every point of the internal boundary;

(*b*) The circulations in the *n* independent circuits of the region have prescribed values; and

(*c*) The velocity vanishes at an infinite distance from the internal boundary.

If, for instance, we have an anchor-ring moving in an infinite mass of liquid which is at rest at infinity, the irrotational motion of the fluid at any instant is determinate when we know the motion of the ring (and therefore the velocity of every element of its surface normal to itself), and also the value of the circulation in any circuit embracing it.

63. The theory of multiple continuity seems to have been first developed by Riemann[*], for spaces of two dimensions, *à propos* of his researches on the theory of functions of a complex variable, in which connection also cyclic functions satisfying the equation

$$\frac{d^2\phi}{dx^2} + \frac{d^2\phi}{dy^2} = 0$$

through multiply-connected regions present themselves.

The bearing of the theory on Hydrodynamics, and the existence in certain cases of many-valued velocity-potentials were first pointed out by Helmholtz[†]. The subject of cyclic irrotational motion in multiply-connected regions was afterwards taken up and fully investigated by Sir W. Thomson in his paper on vortex-motion already referred to.

Green's Theorem.

64. In treatises on Electrostatics, &c., many important properties of the potential are usually proved by means of a certain theorem due to Green[‡]. Of these the most interesting from our present point of view have been already given; but as the theorem

---

[*] Lehrsätze aus der Analysis Situs. *Crelle*, t. 54. 1857.

[†] *Crelle*, t. 55. 1858.

[‡] An essay on Electricity and Magnetism, § 3.

in question leads to a useful expression for the kinetic energy in any case of irrotational motion, we give the following proof of it.

Let u, v, w, ϕ be any functions which are finite, continuous, and single-valued at all points of a connected region S completely bounded by one or more closed surfaces; let dS be an element of any one of these surfaces, l, m, n the direction-cosines of the normal to it drawn inwards.   We shall prove that

$$\iint \phi \, (lu + mv + nw) \, dS$$
$$= - \iiint \left(\frac{d \cdot \phi u}{dx} + \frac{d \cdot \phi v}{dy} + \frac{d \cdot \phi w}{dz} \right) dx \, dy \, dz \ldots (17),$$

where the double-integral is taken over the whole boundary of S, and the triple-integral throughout its interior. ·

If we conceive a series of surfaces drawn so as to divide S into any number of separate parts, the integral

$$\iint \phi \, (lu + mv + nw) \, dS \ldots \ldots \ldots \ldots \ldots (18),$$

taken over the boundary of S, is equal to the sum of the similar integrals taken each over the whole boundary of one of these parts. For, for every element $d\sigma$ of a dividing surface, we have, in the integrals corresponding to the parts lying on the two sides of this surface, elements $\phi \, (lu + mv + nw) \, d\sigma$, and $\phi \, (l'u + m'v + n'w) \, d\sigma$, respectively.   But the normals to which l, m, n, l', m', n' refer being drawn inwards in each case, we have $l' = - l$, $m' = - m$, $n' = - n$; so that in forming the sum of the integrals spoken of the elements due to the dividing surfaces disappear, and we have left only those due to the original boundary of S.

Now let us suppose the dividing surfaces to consist of three infinite series of planes parallel to yz, zx, xy, respectively.   Let x, y, z be the co-ordinates of the centre of one of the rectangular spaces thus formed, dx, dy, dz the lengths of its edges, and let us calculate the value of (18) taken over the boundary of this space. As in Art. 8 the part of the integral due to the yz-face nearest the origin is $\left(\phi u - \frac{1}{2} \frac{d \cdot \phi u}{dx} \, dx \right) dy \, dz$, and that due to the opposite face

is $-\left(\phi u + \frac{1}{2}\frac{d \cdot \phi u}{dx}dx\right)dy\,dz$. The sum of these is $-\frac{d \cdot \phi u}{dx}dx\,dy\,dz$.
Calculating in the same way the parts of the integral due to the
remaining pairs of faces, we get for the final result

$$\dot{-}\left(\frac{d \cdot \phi u}{dx} + \frac{d \cdot \phi v}{dy} + \frac{d \cdot \phi w}{dz}\right)dx\,dy\,dz \,\ldots\ldots\ldots\, (19).$$

Hence (17) simply expresses the fact that the surface-integral (18),
taken over the boundary of S, is equal to the sum of the similar
integrals taken over the boundaries of the elementary spaces of
which we have supposed S built up.

We may interpret (17) by regarding u, v, w as the component
velocities of a continuous system of points filling the region S, and
supposing ϕ to represent some property (estimated per unit
volume) which they carry with them in their motion. The surface-
integral on the left-hand side of (17) expresses then the amount
of ϕ which enters S in unit time across its boundary; whilst the
above investigation shews that (19) expresses the rate at which
the property ϕ is being accumulated in the elementary space
$dx\,dy\,dz$. The theorem then asserts that the total increase of ϕ
within the region is equal to the influx across the boundary. A
particular case is where u, v, w are the component velocities of a
fluid filling the region, and ϕ is put $= \rho$, the density. See Art. 12.

Corollary 1. Let $\phi = 1$; the theorem becomes

$$\iint (lu + mv + nw)\,dS = -\iiint \left(\frac{du}{dx} + \frac{dv}{dy} + \frac{dw}{dz}\right)dx\,dy\,dz.$$

If u, v, w be the component velocities of a liquid filling the region
S, the right-hand side of this equation vanishes, by the equation
of continuity; so that

$$\iint (l\dot{u} + mv + nw)\,dS = 0 \,\ldots\ldots\ldots\ldots\ldots\, (20),$$

which expresses that as much fluid leaves the region as enters it.

Corollary 2. Let u, v, $w = \frac{d\psi}{dx}$, $\frac{d\psi}{dy}$, $\frac{d\psi}{dz}$, respectively, where
ψ is a function which, with its first differential coefficients, is
finite and continuous throughout S. Then

$$lu + mv + nw = \frac{d\psi}{dn},$$

where dn is an element of the inwardly-directed normal to the surface of S. Substituting in (17) and performing the differentiations indicated, we find

$$\iint \phi \frac{d\psi}{dn} dS = - \iiint \left(\frac{d\phi}{dx} \frac{d\psi}{dx} + \frac{d\phi}{dy} \frac{d\psi}{dy} + \frac{d\phi}{dz} \frac{d\psi}{dz} \right) dx\,dy\,dz$$

$$- \iiint \phi \nabla^2 \psi\, dx\,dy\,dz \ldots\ldots\ldots\ldots\ldots\ldots(21).$$

By simply interchanging ϕ and ψ we obtain (provided ψ be single-valued),

$$\iint \psi \frac{d\phi}{dn} dS = - \iiint \left(\frac{d\phi}{dx} \frac{d\psi}{dx} + \frac{d\phi}{dy} \frac{d\psi}{dy} + \frac{d\phi}{dz} \frac{d\psi}{dz} \right) dx\,dy\,dz$$

$$- \iiint \psi \nabla^2 \phi\, dx\,dy\,dz \ldots\ldots\ldots\ldots\ldots\ldots(22).$$

Equations (21) and (22) together constitute Green's theorem.

Corollary 3. In (21) let ϕ be the velocity-potential of a liquid, and let $\psi = 1$; we find, since $\nabla^2 \phi = 0$,

$$\iint \frac{d\phi}{dn} dS = 0 \ldots\ldots\ldots\ldots\ldots\ldots\ldots(23),$$

which is in fact what (20) becomes for the case of irrotational motion. Compare Art. 44.

Corollary 4. In (21) let $\psi = \phi$, and let ϕ be the velocity-potential of a liquid. We obtain

$$\iiint \left\{ \left(\frac{d\phi}{dx} \right)^2 + \left(\frac{d\phi}{dy} \right)^2 + \left(\frac{d\phi}{dz} \right)^2 \right\} dx\,dy\,dz = - \iint \phi \frac{d\phi}{dn} dS \ldots\ldots(24).$$

If we multiply this equation by $\frac{1}{2}\rho$ it becomes susceptible of a simple dynamical interpretation. On the right-hand side $\frac{d\phi}{dn}$ denotes the normal velocity of the fluid inwards, whilst $-\rho\phi$ is, by Art. 26, the impulsive pressure necessary to generate the actual motion. It is a proposition in Dynamics* that the work done by an impulse is measured by the product of the impulse into half the sum of the initial and final velocities, resolved in the direction of the impulse, of the point to which it is applied. Hence the right-hand side of (24) when modified as described, expresses the work done by the system of impulsive pressures which, applied to to the surface of S, generate the actual motion; whilst the left-hand side gives the kinetic energy of this motion. The equation (24) asserts that these two quantities are equal, thus verifying for our particular case the principle of energy.

Hence if T denote the total kinetic energy of the liquid, we have

$$2T = -\rho \iint \phi \frac{d\phi}{dn} dS \dots\dots\dots\dots\dots(25).$$

Corollary 5. In (24) instead of ϕ let us write $\frac{d\phi}{dx}$, which will of course satisfy (10) as ϕ does; and let us apply the resulting theorem to the region included within a spherical surface of radius r having any point (x, y, z) as centre. With the same notation as in Art. 46, we have

$$\tfrac{1}{2} r^2 \frac{d}{dr} \iint u^2 d\varpi = \iint u \frac{du}{dr} dS = -\iint \frac{d\phi}{dx} \frac{d}{dn} \cdot \frac{d\phi}{dx} dS$$

$$= \iiint \left\{ \left(\frac{d^2\phi}{dx^2}\right)^2 + \left(\frac{d^2\phi}{dx\,dy}\right)^2 + \left(\frac{d^2\phi}{dx\,dz}\right)^2 \right\} dx\,dy\,dz,$$

an essentially positive quantity. Hence, writing $q^2 = u^2 + v^2 + w^2$, we see that

$$\frac{d}{dr} \iint q^2 d\varpi$$

is positive; *i.e.* the mean value of q^2, taken over a sphere having any point as centre, increases with the radius of the sphere. Hence q cannot be a maximum at any point of the fluid, as was otherwise proved in Art. 45.

65. We shall require to know, hereafter, the form assumed by the expression (25) for the kinetic energy when the fluid extends

---

* Thomson and Tait, *Natural Philosophy*, Art. 308.

to infinity and is at rest there, being limited internally by one or
more closed surfaces S.   Let us suppose a large closed surface Σ
described so as to enclose the whole of S.   The energy of the fluid
included between S and Σ is

$$-\tfrac{1}{2}\rho \iint \phi \frac{d\phi}{dn} dS - \tfrac{1}{2}\rho \iint \phi \frac{d\phi}{dn} d\Sigma \dots\dots\dots\dots\dots(26),$$

where the integration in the first term extends over S, that in the
second over Σ.   Since we have by (11)

$$\iint \frac{d\phi}{dn} dS + \iint \frac{d\phi}{dn} d\Sigma = 0,$$

(26) may be written

$$-\tfrac{1}{2}\rho \iint (\phi - C) \frac{d\phi}{dn} dS - \tfrac{1}{2}\rho \iint (\phi - C) \frac{d\phi}{dn} d\Sigma \dots\dots\dots(27),$$

where C may be any constant, but is here supposed to be the
constant value to which ϕ was shewn in Art. 51 to tend at an
infinite distance from S.   Now the whole region occupied by the
fluid may be supposed made up of tubes of flow, each of which
must pass either from one point of the internal boundary to another, or from that boundary to infinity.   Hence the value of the
integral

$$\iint \frac{d\phi}{dn} d\Sigma,$$

taken over any surface, open or closed, finite or infinite, drawn
within the region, must be finite.   Hence ultimately, when Σ is
taken infinitely large and infinitely distant all round from S, the
second term of (27) vanishes, and we have

$$2T = -\rho \iint (\phi - C) \frac{d\phi}{dn} dS \dots\dots\dots\dots\dots(28),$$

where the integration extends over the internal boundary only.

If the total flux across the internal boundary be zero, we have

$$\iint \frac{d\phi}{dn} dS = 0,$$

so that (28) becomes

$$2T = -\rho \iint \phi \frac{d\phi}{dn} dS \dots\dots\dots\dots\dots(29),$$

simply.

Thomson's Extension of Green's Theorem.

66. It was assumed in the proof of Green's Theorem that ϕ and ψ were both single-valued functions. If either be a cyclic function, as may be the case when the region to which the integrations in (21) and (22) refer is multiply-connected, the statement of the theorem must be modified. Let us suppose, for instance, that ϕ is cyclic; the surface-integral on the left-hand side of (21), and the second volume-integral on the right-hand side, are then indeterminate, on account of the ambiguity in the value of ϕ itself. To remove this ambiguity, let the barriers necessary to reduce the region to a simply-connected one be drawn, as explained in Art. 54. We may suppose such values assigned to ϕ that it shall be continuous and single-valued throughout the region thus modified (Art. 56); and equation (21) will then hold, provided the two sides of each barrier be reckoned as part of the boundary of the region, and therefore included in the surface-integral on the left-hand side. Let $d\sigma_1$ be an element of one of the barriers, κ_1 the cyclic constant corresponding to that barrier, $\dfrac{d\psi}{dn}$ the rate of variation of ψ in the positive direction of the normal to $d\sigma_1$. Since, in the parts of the surface-integral due to the two sides of $d\sigma_1$, $\dfrac{d\psi}{dn}$ is to be taken with opposite signs, whilst the value of ϕ on the negative side exceeds that on the positive side by κ_1, we get finally for the element of the integral due to $d\sigma_1$, the value $-\kappa_1 \dfrac{d\psi}{dn} d\sigma_1$. Hence (21) becomes, in the altered circumstances,

$$\iint \phi \frac{d\psi}{dn}\, dS - \kappa_1 \iint \frac{d\psi}{dn} d\sigma_1 - \kappa_2 \iint \frac{d\psi}{dn} d\sigma_2 - \&c.$$

$$= -\iiint \left(\frac{d\phi}{dx}\frac{d\psi}{dx} + \frac{d\phi}{dy}\frac{d\psi}{dy} + \frac{d\phi}{dz}\frac{d\psi}{dz} \right) dx\,dy\,dz$$

$$- \iiint \phi \nabla^2 \psi\, dx\,dy\,dz \dots\dots\dots\dots\dots\dots\dots\dots(30);$$

where the surface-integrations indicated on the left-hand side extend, the first over the original boundary of the region only, and the rest over the several barriers. The coefficient of any κ is evidently the total flux across the corresponding barrier, in a motion of which ψ is the velocity-potential. The values of ϕ in the first and last terms of the equation are to be assigned in the manner indicated in Art. 56.

If ψ also be a cyclic function, having the cyclic constants κ_1', κ_2', &c., then (22) becomes in the same way

$$\iint \psi \frac{d\phi}{dn} \, dS - \kappa_1' \iint \frac{d\phi}{dn} \, d\sigma_1 - \kappa_2' \iint \frac{d\phi}{dn} \, d\sigma_2 - \&\text{c.}$$

$$= -\iiint \left(\frac{d\phi}{dx} \frac{d\psi}{dx} + \frac{d\phi}{dy} \frac{d\psi}{dy} + \frac{d\phi}{dz} \frac{d\psi}{dz} \right) dx\, dy\, dz$$

$$- \iiint \psi \nabla^2 \phi \, dx\, dy\, dz \dots\dots\dots\dots\dots\dots\dots(31).$$

Equations (30) and (31) together constitute Thomson's extension of Green's theorem.

67. If in (30) we put $\psi = \phi$, and suppose ϕ to be the velocity-potential of an incompressible fluid, we find

$$2T = \rho \iiint \left\{ \left(\frac{d\phi}{dx}\right)^2 + \left(\frac{d\phi}{dy}\right)^2 + \left(\frac{d\phi}{dz}\right)^2 \right\} dx\, dy\, dz$$

$$= -\rho \iint \phi \frac{d\phi}{dn} \, dS + \rho\kappa_1 \iint \frac{d\phi}{dn} \, d\sigma_1 + \rho\kappa_2 \iint \frac{d\phi}{dn} \, d\sigma_2 + \&\text{c.} \dots(32).$$

To interpret the last member of this formula we must recur to the artificial method of generating cyclic irrotational motion explained in Art. 61. The first term has already been interpreted as twice the work done by the impulsive pressure $-\rho\phi$ applied to every part of the original boundary of the fluid. Again, $\rho\kappa_1$ is the impulsive pressure applied, in the positive direction, to the infinitely thin massless membrane by which the place of the first barrier was in Art. 61 supposed to be occupied; so that the expression $\frac{1}{2} \iint \rho\kappa_1 . \frac{d\phi}{dn} d\sigma_1$ denotes the work done by the impulsive forces applied to that membrane; and so on. Hence (32) expresses the fact that the energy of the motion is equal to the work done by the whole system of impulsive forces by which we may suppose it generated.

In applying (32) to the case where the fluid extends to infinity and is at rest there, we must replace the first term of the third member by

$$-\rho \iint (\phi - C)\, \frac{d\phi}{dn}\, dS \dots\dots\dots\dots (33),$$

where the integration extends over the internal boundary only; or, when the total flux across this boundary is zero, by

$$-\rho \iint \phi\, \frac{d\phi}{dn}\, dS \dots\dots\dots\dots (34).$$

The proof is the same as in Art. 65.

CHAPTER IV.

MOTION OF A LIQUID IN TWO DIMENSIONS.

68. IF the velocities u, v be functions of x, y only, whilst w is zero, the motion takes place in a series of planes parallel to xy, and is the same in each of those planes. The investigation of the motion of a liquid under these circumstances is characterized by certain analytical peculiarities; and the solutions of several problems of great interest are readily obtained.

Since the whole motion is known when we know that in the plane $z = 0$, we shall confine our attention to the motion which takes place in that plane. When we speak of points and lines drawn in that plane, we shall in general understand them to represent respectively the straight lines parallel to the axis of z, and the cylindrical surfaces having their generating lines parallel to the axis of z, of which they are the traces.

By the flux across any curve we shall understand the volume of fluid which in unit time crosses that portion of the cylindrical surface having the curve as base, which is included between the planes $z = 0$, $z = 1$.

69. Let A, P be any two points in the plane xy. The flux across any two lines joining AP is the same, provided they can be reconciled without passing out of the region occupied by the moving liquid; for otherwise the space included between these two lines would be gaining or losing fluid. Hence if A be fixed, and P variable, the flux across any line AP is a

function of the position of P. Let ψ be the function; more precisely, let ψ denote the flux across AP *from left to right,* as regards an observer placed on the curve, and looking along it from A in the direction of P. Analytically, if l, m be the direction-cosines of the normal (drawn to the right) to any element ds of the curve, we have

$$\psi = \int_A^P (lu + mv)\, ds \dots\dots\dots\dots\dots\dots(1).$$

If the region occupied by the liquid be aperiphractic, ψ is necessarily a single-valued function, but in periphractic regions the value of ψ may depend on the nature of the path joining AP. For spaces of two dimensions however periphraxy and multiple-continuity become the same thing, so that the properties of ψ, when it is a many-valued function, in relation to the nature of the region occupied by the moving liquid, may be inferred from Arts. 55, 56, where we have discussed the same question with regard to ϕ.

The cyclic constants of ψ, when the region is periphractic, are the values of the flux across the closed curves forming the several parts of the internal boundary.

· A change, say from A to B, of the point from which ψ is reckoned has merely the effect of adding a constant, viz. the flux across a line BA, to the value of ψ; so that we may, if we please, regard ψ as indeterminate to the extent of an arbitrary constant.

If P move about in such a manner that the value of ψ does not alter, it will trace out a curve such that no fluid anywhere crosses it, *i.e.* a stream-line. Hence the curves $\psi = $ const. are the stream-lines, and ψ is called the 'stream-function.'

If P receive an infinitesimal displacement $PQ (= dy)$ parallel to y, the increment of ψ is the flux across PQ from left to right, *i.e.* $d\psi = u \cdot PQ$, or

$$u = \frac{d\psi}{dy} \dots\dots\dots\dots\dots\dots (2).$$

Again displacing P parallel to x, we find in the same way

$$v = -\frac{d\psi}{dx} \dots\dots\dots\dots\dots\dots (3).$$

The existence of a function ψ related to u and v in this manner might also have been inferred from the form which the equation of continuity takes in this case, viz.

$$\frac{du}{dx} + \frac{dv}{dy} = 0 \dots\dots\dots (4),$$

which is the analytical condition that $u\,dy - v\,dx$ should be an exact differential.

The foregoing considerations apply whether the motion be rotational or irrotational. The formulæ for the component angular velocities, given in Art. 38, become

$$\xi = 0, \quad \eta = 0, \quad \zeta = -\tfrac{1}{2}\left(\frac{d^2\psi}{dx^2} + \frac{d^2\psi}{dy^2}\right) \dots\dots\dots (5);$$

so that in irrotational motion we have

$$\frac{d^2\psi}{dx^2} + \frac{d^2\psi}{dy^2} = 0 \dots\dots\dots (5a).$$

70. In what follows we confine ourselves to the case of irrotational motion, which is, as we have already seen, characterized by the existence, in addition, of a velocity-potential ϕ, connected with u, v by the relations

$$u = \frac{d\phi}{dx}, \qquad v = \frac{d\phi}{dy} \dots\dots\dots (6),$$

and, since we are considering the motion of incompressible fluids only, satisfying the equation of continuity

$$\frac{d^2\phi}{dx^2} + \frac{d^2\phi}{dy^2} = 0 \dots\dots\dots (7).$$

The theory of the function ϕ, and the connection of its properties with the nature of the two-dimensional space through which the irrotational motion holds, may be readily inferred from the corresponding theorems in three dimensions proved in the last chapter. The alterations, both in the enunciation and in the proof, which are requisite to adapt these to the case of two dimensions are for the most part purely verbal. An exception, which we will briefly examine, occurs however in the case of the theorem of Art. 46 and of those which depend on it.

71. If ds be an element of the boundary of any portion of the plane xy which is occupied wholly by moving liquid, and if dr be an element of the normal to ds drawn inwards, we have, by Art. 44,

$$\int \frac{d\phi}{dr} ds = 0 \quad \dotfill (8),$$

the integration extending round the whole boundary. If this boundary be a circle, and if r, θ be polar co-ordinates referred to the centre P of this circle as origin, the equation (8) may be written

$$\int_0^{2\pi} \frac{d\phi}{dr} \cdot r d\theta = 0,$$

or

$$\frac{d}{dr} \int_0^{2\pi} \phi d\theta = 0.$$

Hence the value of the integral $\dfrac{1}{2\pi} \displaystyle\int_0^{2\pi} \phi d\theta$, *i.e.* the mean-value of ϕ over a circle of centre P, and radius r, is independent of the value of r, and therefore remains unaltered when r is diminished without limit, in which case it becomes the value of ϕ at P.

If the region occupied by the fluid be periphractic, and we apply (8) to the space enclosed between one of the internal boundaries and a circle with centre P and radius r surrounding this boundary, and lying wholly in the fluid, we have

$$\int_0^{2\pi} \frac{d\phi}{dr} \cdot r d\theta = 2\pi M;$$

where the integration in the first member extends over the circle only, and $2\pi M$ denotes the flux into the region across the internal boundary. Hence

$$\frac{d}{dr} \cdot \frac{1}{2\pi} \int_0^{2\pi} \phi d\theta = \frac{M}{r},$$

which gives on integration

$$\frac{1}{2\pi} \int_0^{2\pi} \phi d\theta = M \log r + C \dotfill (9);$$

i.e. the mean value of ϕ over a circle with centre P and radius r is equal to $M \log r + C$, where C is independent of r but may vary with the position of P. This formula holds of course only so long as the circle embraces the same internal boundary, and lies itself wholly in the fluid.

If the region be unlimited externally, and if the circle embrace the whole of the internal boundaries, and if further the velocity be everywhere zero at infinity, then C is an absolute constant; as is seen by reasoning similar to that of Art. 46.

It may then be shewn, exactly as in Art. 51, that the value of ϕ at a very great distance r from the internal boundary tends to the value $M \log r + C$. In the particular case when $M = 0$ the limit to which ϕ tends at infinity is finite; in all other cases it is infinite, and of the same sign as M.

We infer, as in Art. 52, that there is only one single-valued function ϕ which (a) satisfies the equation (7) at every point of the plane xy external to a given system of closed curves, (b) makes the value of $\dfrac{d\phi}{dn}$ equal to an arbitrarily given quantity at every point of these curves, and (c) has its first differential coefficients all zero at infinity.

72.   The kinetic energy of a portion of fluid bounded by a cylindrical surface whose generating lines are parallel to the axis of z, and two planes perpendicular to the axis of z at unit distance, is given by the formula

$$\tfrac{1}{2}\rho \iint \left\{ \left(\frac{d\phi}{dx}\right)^2 + \left(\frac{d\phi}{dy}\right)^2 \right\} \, dx\,dy = -\tfrac{1}{2}\rho \int \phi \frac{d\phi}{dn} \, ds \ldots \ldots (10),$$

where the surface-integral is taken over the portion of the plane xy cut off by the cylindrical surface, and the line-integral round the boundary of this portion.

If the cylindrical part of the boundary consist of two or more separate portions one of which embraces all the rest, the enclosed region is multiply-connected, and the equation (10) needs a correction, which may be applied exactly as in Art. 66.

If we attempt, by a process similar to that of Art. 65, to calculate the energy in the case where the region extends to infinity, we find that its value is infinite, except when $M = 0$.

For if we introduce a circle of great radius r as the external boundary of the portion of the plane xy considered, we find that the corresponding part of the integral on the right-hand side of (10) tends, as r increases, to the value $\pi\rho M(M\log r + C)$, and is therefore ultimately infinite. The only exception is when $M = 0$, in which case we may suppose the line-integral in (10) to extend over the internal boundary only.

73.   The functions ϕ and ψ are connected by the relations

$$\frac{d\phi}{dx} = \frac{d\psi}{dy}, \qquad \frac{d\phi}{dy} = -\frac{d\psi}{dx} \dots\dots\dots\dots(11).$$

These are the conditions that $\phi + i\psi$, where i stands for $\sqrt{-1}$, should be a function of the 'complex' variable $x + iy$. For if

$$\phi + i\psi = f(x + iy)\dots\dots\dots\dots\dots(12),$$

we have

$$\frac{d}{dy}(\phi + i\psi) = if'(x + iy) = i\frac{d}{dx}(\phi + i\psi)\dots\dots\dots(13),$$

whence, equating separately the real and the imaginary parts, we obtain (11).

Hence any assumption of the form (12) gives a possible case of irrotational motion. The curves $\phi = $ const. are the curves of equal velocity-potential, and the curves $\psi = $ const. are the stream-lines. Since, by (11),

$$\frac{d\phi}{dx}\frac{d\psi}{dx} + \frac{d\phi}{dy}\frac{d\psi}{dy} = 0,$$

we see that these two systems of curves cut one another at right angles, as we have already proved. See Art. 27. Since the relations (11) are unaltered when we write $-\psi$ for ϕ, and ϕ for ψ, we may, if we choose, look upon the curves $\psi = $ const. as the equipotential curves, and the curves $\phi = $ const. as the stream-lines; so that every assumption of the kind indicated gives us *two* possible cases of irrotational motion.

74.   The fundamental property of a function of a complex variable, from which all others flow, is that it has a differential coefficient with respect to that variable. If ϕ, ψ denote any functions whatever of x and y, then corresponding to every value

of $x + iy$ there must be one or more definite values of $\phi + i\psi$; but the ratio of the differential of this function to that of $x + iy$, viz.

$$\frac{d\phi + id\psi}{dx + idy},$$

or

$$\frac{\left(\frac{d\phi}{dx} + i\frac{d\psi}{dx}\right) dx + \left(\frac{d\phi}{dy} + i\frac{d\psi}{dy}\right) dy}{dx + idy},$$

depends in general on the ratio $dx : dy$. The condition that it should be the same for all values of this ratio is

$$\frac{d\phi}{dy} + i\frac{d\psi}{dy} = i\left(\frac{d\phi}{dx} + i\frac{d\psi}{dx}\right) \ldots\ldots\ldots\ldots(13),$$

which is equivalent to (11).

We may therefore take this property as the definition of a function of the complex variable $x + iy$; viz. such a function must have, for every assigned value of the variable, not only a definite value or system of values, but also for each of these values a definite differential coefficient. The advantage of this definition is that it is quite independent of the existence of an analytical expression for the function.

The theory of functions of this kind has received considerable development at the hands of Cauchy, Riemann, and others; and has grown into an important branch of mathematical analysis. We give here only such elementary notions connected with the subject as are of immediate hydrodynamical interest*.

75. We assume the student to be acquainted with the method of representing the symbol $x + iy$ by a vector drawn from the origin of rectangular co-ordinates to the point (x, y).

In this method the sum of two vectors is defined to be the vector drawn from the origin to the opposite corner of the parallelogram of which they form adjacent sides.

The effect of multiplying one vector $x + iy$ by another $a + ib$ is to increase its length in the ratio $r : 1$, and to turn it in the

---

* The reader who wishes for an elementary exposition of the analytical theory may consult Durège: *Elemente der Theorie der Functionen einer complexen veränderlichen Grösse.* 2nd ed., Leipzig, 1873.

positive direction (*i.e.* from x to y) through an angle θ, where $r = \surd(a^2 + b^2)$, and θ is the least positive value of arc $\tan\dfrac{b}{a}$. With respect to the expression $a + ib$, r is called the 'modulus,' and θ the 'amplitude.'

The meanings of subtraction and division of vectors follow at once from the considerations that they are the operations inverse to those of addition and multiplication, respectively.

With these conventions, the addition, multiplication, &c., of vectors are performed according to the same laws of operation as in common algebra.

76.   For shortness we denote the complex quantities $\phi + i\psi$, and $x + iy$ by the letters w, and z, respectively. These symbols not being required at present in their former meanings may without inconvenience have these new ones assigned to them. Then w being any function of z, according to the definition of Art. 74, we have corresponding to any point P of the plane xy (which we may call the plane of the variable z) one or more definite values of w. Let us choose any one of these, and denote it by a point P' of which ϕ, ψ are the rectangular co-ordinates in a second plane (the plane of the function w). If P trace out any curve in the plane of z, P' will trace out a corresponding curve in the plane of w. By mapping out the positions of P' corresponding to the various points P of the plane xy, we may exhibit graphically all the properties of the function w.

Let now Q be a point infinitely near to P, and let Q' be the corresponding point infinitely near to P'. We may denote PQ by dz, $P'Q'$ by dw. The vector $P'Q'$ may be obtained from the vector PQ by multiplying it by the differential coefficient $\dfrac{dw}{dz}$, whose value is by definition dependent only on the position of P, and not on the direction of the element dz (PQ). Now the effect of multiplying any vector by the complex quantity $\dfrac{dw}{dz}$ is to increase its length in some definite ratio, and to turn it in the positive direction through some definite angle. Hence, in the transition from the plane of z to that of w, all the infinitesimal vectors drawn from the point P have their lengths altered in the same ratio, and are turned through the same angle. Hence any angle in the plane of z is equal to the corresponding angle in the plane of w, and any infinitely small figure in the one plane is similar to the corresponding figure in the other. In other words, corresponding figures in the planes of z and w are similar in their infinitely small parts.

For instance, in the plane of w the straight lines $\phi = \text{const.}$, $\psi = \text{const.}$, where the constants have assigned to them a series of values in arithmetical progression, the common difference being infinitesimal and the same in each case, form two systems of straight lines at right angles dividing the plane into infinitely small squares. Hence in the plane xy the corresponding curves $\phi = \text{const.}$, $\psi = \text{const.}$, the values of the constants being assigned to them as before cut one another at right angles (as has already been proved otherwise) and divide the plane into a series of infinitely small squares.

The similarity of corresponding infinitely small portions of the planes w and z breaks down at points where the differential coefficient $\dfrac{dw}{dz}$ is zero or infinite. Since $\dfrac{dw}{dz} = \dfrac{d\phi}{dx} + i\dfrac{d\psi}{dx}$, the corresponding value of the velocity, in the hydrodynamical application, is zero or infinity.

77. The processes of differentiation and integration, as applied to functions of a complex variable, claim a little notice.

The conditions (13) that w should be a function of z may be written

$$\frac{dw}{dy} = i\frac{dw}{dx}\dots\dots\dots\dots\dots\dots(14),$$

a form which the student should interpret. It is obvious that (14) is satisfied when $\dfrac{dw}{dz}\left[= \dfrac{dw}{dx}\right]$ is written for w, and therefore that $\dfrac{dw}{dz}$ is itself a function of z. Hence all the derivatives of a function of a complex variable are themselves functions of that variable.

The meaning of the integral

$$\int_{z_0}^{z} w\,dz \dots\dots\dots\dots\dots\dots (15),$$

taken *along any assigned path* from z_0 to z, is defined as follows. Supposing the path divided into infinitesimal portions, we form the product $w\,dz$ for each of them, and add the results. It is easily shewn that the value of the integral is to a certain extent independent of the nature of the path joining z_0 and z. The theorem of Art. 40, when applied to a plane space (xy) becomes

$$\iint\left(\frac{dv}{dx} - \frac{du}{dy}\right)dx\,dy = \int(u\,dx + v\,dy) \dots\dots\dots (16).$$

The double integral extends over any portion of the plane xy throughout which u and v are finite and continuous, and their first derivatives finite; whilst the single integral extends in the positive direction (see Art. 39) round the boundary of that portion. Now if we write $u = w$, $v = iw$, it follows by (14) that $\int(w\,dx + iw\,dy)$, or $\int w\,dz$, is zero when taken round the boundary of any portion of the plane xy throughout which w is finite and continuous, and its first derivative finite.

We infer, as in Art. 41, that the integral (15) is the same for any two paths joining z, z_0, so long as these paths do not include between them any points at which w is infinite or discontinuous, or $\frac{dw}{dz}$ infinite.

Any points at which these conditions are violated may be isolated by drawing an infinitely small closed curve around each. The rest of the plane xy then forms a multiply-connected region. The value of the integral $\int w\,dz$ taken round any evanescible circuit drawn in this region is zero; and the integral (15) is the same for any two reconcileable paths. The values of (15) corresponding to two irreconcileable paths differ by a quantity of the form $p_1\kappa_1 + p_2\kappa_2 + \dots\dots$, where p_1, p_2,...... are integers, and κ_1, κ_2,...... denote the values of $\int w\,dz$ taken round the several circuits surrounding the above-mentioned points*. It is unnecessary to

---

* In the analytical theory κ_1, κ_2,... are called the 'moduli of periodicity' of the integral (15).

dwell on the proof of these statements, or to enter more fully into the theory of many-valued integrals of the form (15); to do so would be to repeat, with merely verbal alterations, portions of Chapter III.

The integral (15) is itself a function of z, according to the definition of Art. 74. For, denoting its value by Z, we have,

obviously $$\frac{dZ}{dz} = w,$$

i.e. Z has a definite differential coefficient.

78.   An important illustration of the above theory is furnished by the integral

$$\int \frac{dz}{z} \quad\dots\dots\dots\dots\dots\dots\dots (17).$$

The only point at which the function z^{-1}, or its derivative, is infinite or discontinuous, is the origin. Introducing polar co-ordinates, we write

$$z = r(\cos\theta + i\sin\theta) \dots\dots\dots\dots\dots (18),$$

whence $$\frac{dz}{z} = \frac{dr}{r} + id\theta \dots\dots\dots\dots\dots (19).$$

Hence the value of (17) taken round an infinitely small circle having the origin as centre is

$$\int \frac{dz}{z} = \int id\theta = 2\pi i.$$

In the analytical theory above referred to, the logarithm of a complex quantity z is defined by the equation

$$\log z = \int_1^z \frac{dz}{z} \dots\dots\dots\dots\dots (20).$$

Hence $\log z$ is a many-valued function, the cyclic constant being $2\pi i$.

The properties of the logarithmic function readily follow from the definition (20). Thus if z_1, z_2 be any two complex quantities we have

$$\frac{d \cdot z_1 z_2}{z_1 z_2} = \frac{dz_1}{z_1} + \frac{dz_2}{z_2},$$

and therefore

$$\int_1^{z_1 z_2} \frac{dz}{z} = \int_1^{z_1} \frac{dz_1}{z_1} + \int_1^{z_2} \frac{dz_2}{z_2},$$

or
$$\log (z_1 z_2) = \log z_1 + \log z_2 \dots\dots\dots\dots(21).$$

Putting $z_2 = 0$, we have

$$\log 0 = \log z_1 + \log 0,$$

whence
$$\log 0 = \infty,$$

or rather, since it appears that the real part of the logarithm of a small quantity is essentially negative,

$$\log 0 = -\infty \dots\dots\dots\dots\dots(22).$$

In the same way

$$\log \infty = \infty \dots\dots\dots\dots\dots (23).$$

Let us examine the properties of the function 'inverse' to the logarithm; viz. writing

$$w = \log z,$$

let us investigate the properties of z as a function of w. This function we denote by e^w. It follows at once from (21) that

$$e^{w_1} . e^{w_2} = e^{w_1 + w_2} \dots\dots\dots\dots (24),$$

the fundamental property of the 'exponential' function. Hence, and from (22), (23),

$$e^0 = 1, \quad e^{+\infty} = +\infty, \quad e^{-\infty} = 0.$$

Also since $\log z$ is cyclic, the constant being $2\pi i$, e^w is a *periodic* function, the period being $2\pi i$, viz.

$$e^{w \pm 2n\pi i} = e^w,$$

where n is any integer.

Let us map out, in the manner explained in Art. 76, the relation between the two functions z and w. It appears from (19) that

$$w = \log z = \int_1^r \frac{dr}{r} + i\theta \dots\dots\dots\dots(25)^*.$$

The first term on the right-hand side of (25) is essentially real; we denote it by $\log r$. We have then $\log r$ and θ as the rect-

---

* Putting $r = 1$ in (18) and (25) we see that
$$e^{i\theta} = \cos \theta + i \sin \theta.$$

angular co-ordinates of the point P' in the plane of w corresponding to the point $P\,(x, y)$ of the plane of z. If P describe a circle of

Fig. 5.

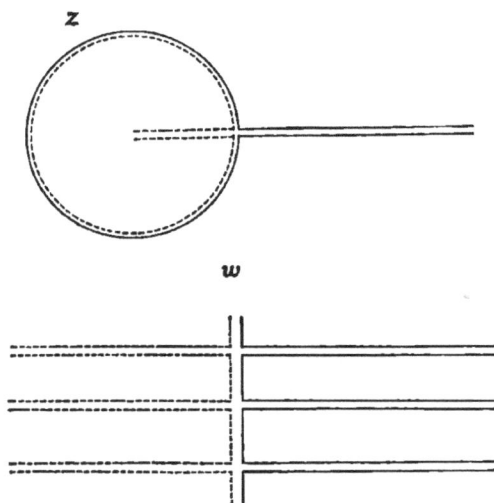

radius unity about the origin, starting from the point $(1, 0)$; then, since $\log 1 = 0$, P' will move along the axis of ψ, and will describe a length 2π of that axis for every revolution of P'. To points P outside the circle $r = 1$ correspond points P' to the right of the axis of ψ; and *vice versa*. To every straight line through the origin, in the plane of z, corresponds in the plane of w a straight line parallel to the axis of ϕ. The periodicity of z, and the cyclosis of w, are manifested by the division of the plane of w by the straight lines $\psi = 2n\pi$ into an infinite number of compartments, each of which corresponds to the *whole* of the plane of z.

In the same way the properties of the function arc tan z may be deduced from the definition

$$\text{arc tan } z = \int_0^z \frac{dz}{1 + z^2} = \frac{1}{2i} \log \frac{z - i}{z + i},$$

and, though with much greater difficulty, those of the function arc sin z from the definition

$$\text{arc sin } z = \int_0^z \frac{dz}{\sqrt{1 - z^2}}.$$

79. The hydrodynamical interpretation of cases where w is, in the foregoing sense, a many-valued function of z is obvious from the preceding chapter. It is possible however for w to be ambiguous in another way. Let us take for instance the function

$$w = \sqrt{z} = r^{\frac{1}{2}} (\cos \theta + i \sin \theta)^{\frac{1}{2}} \dots\dots\dots\dots (26).$$

If we start from the point $r = 1$, $\theta = 0$, with the value $w = 1$, the value of the function at any other point is

$$w = r^{\frac{1}{2}} (\cos \tfrac{1}{2}\theta + i \sin \tfrac{1}{2}\theta),$$

where θ must of course be supposed to vary continuously. Hence if the point $P(r, \theta)$ describe a closed curve not embracing the origin, w will return to its former value; but if the path of P encompass the origin this will not be the case; if the motion of P be in the positive direction θ will have increased by 2π, and we shall have $w = -1$. A second circuit round the origin will restore to w its original value. Hence to every point P in the plane of z correspond two values of w, which however pass into one another continuously as P describes a closed curve about the origin, where the two values coincide.

Again, if

$$w = z^{\frac{1}{3}} \dots\dots\dots\dots\dots\dots (27),$$

and we start from any point A with the value w_0, the value of w at A after one circuit of P round the origin will be αw_0, after a second circuit $\alpha^2 w_0$, and after a third $\alpha^3 w_0$, where α is a cube root of unity. Hence to every point P of the plane of z correspond three values of w, forming a cycle which recurs at every third circuit of P round the origin.

A point, such as the origin in the above examples, at which two or more values of a function coincide, is called a 'branch-point'. The similarity in their infinitely small parts of the planes of w and z must obviously break down at a branch-point, so that we must have at such points (Art. 76)

$$\frac{dw}{dz} = 0, \text{ or } \infty .$$

The branch-points of a function w are also branch-points of $\dfrac{dw}{dz}$, and *vice versa*, as is easily seen from the meaning of the latter function.

Hence if an assumption of the form (26) or (27) be hydro-dynamically intelligible, the portion of the plane xy occupied by the fluid must not include any branch-points; otherwise the component velocities would be ambiguous at every point of the fluid. Branch-points may however occur on the boundary of this portion, in which case circulation round them is impossible.

80. We can now proceed to some applications of the foregoing theory.

Example 1.   Assume
$$\cdot\, w = A z^n,$$
and suppose A to be real*. Introducing polar co-ordinates r, θ, we have
$$\phi = A r^n \cos n\theta,$$
$$\psi = A r^n \sin n\theta.$$

(a) If $n = 1$, the stream-lines are a system of straight lines parallel to x, and the equipotential curves are a similar system parallel to y. In this case *any* corresponding figures in the planes of w and z are similar, whether they be finite or infinitesimal.

(b) If $n = 2$, the curves $\phi = \text{const.}$ are a system of rectangular hyperbolas having the axes of co-ordinates as their principal axes, and the curves $\psi = \text{const.}$ are a similar system having the co-ordinate axes as asymptotes. The lines $\theta = 0$, $\theta = \frac{1}{2}\pi$ are parts of the same stream-line $\psi = 0$, so that we may take the positive parts of the axes of x, y as fixed boundaries, and thus obtain the case of a fluid in motion in the angle between two perpendicular walls.

(c) If $n = -1$, we get two systems of circles touching the axes of co-ordinates at the origin. Since now $\phi = \dfrac{A \cos \theta}{r}$, the

---

* If A be complex, the curves $\phi = \text{const.}$, $\psi = \text{const.}$ are not altered in form but only in position, being turned round the origin through an angle arc $\tan \beta$, if
$$A = a + i\beta.$$

velocity at the origin is infinite; to avoid a physical absurdity we must suppose the region to which our formulæ apply to be limited internally by a closed curve surrounding the origin.

(d) If $n = -2$, each system of curves is composed of a double system of lemniscates. The axes of the system $\phi = $ const. coincide with x or y; those of the system $\psi = $ const. bisect the angles between these axes.

(e) By properly choosing the value of n we get a case of irrotational motion in which the boundary is composed of two rigid walls inclined at any angle α. The equation of the stream-lines being

$$r^n \sin n\theta = \text{const.},$$

we see that the lines $\theta = 0$, $\theta = \dfrac{\pi}{n}$ are parts of the same stream-line. Hence we have only to put $\dfrac{\pi}{n} = \alpha$, and so obtain the required solution in the form

$$\phi = Ar^{\frac{\pi}{\alpha}} \cos \frac{\pi\theta}{\alpha}, \quad \psi = Ar^{\frac{\pi}{\alpha}} \sin \frac{\pi\theta}{\alpha}.$$

The component velocities along and perpendicular to r, are

$$A \frac{\pi}{\alpha} r^{\frac{\pi}{\alpha}-1} \cos \frac{\pi\theta}{\alpha}, \text{ and } -A \frac{\pi}{\alpha} r^{\frac{\pi}{\alpha}-1} \sin \frac{\pi\theta}{\alpha};$$

and are therefore zero, finite, or infinite at the origin, according as α is less than, equal to, or greater than π.

81. *Example* 2. The assumption

$$w = \mu \log z,$$

or

$$\phi + i\psi = \mu \log re^{i\theta},$$

gives

$$\phi = \mu \log r, \quad \psi = \mu\theta.$$

The equipotential lines are concentric circles about the origin; the stream-lines are straight lines radiating from the origin. Or, we may take the circles $r = $ const. as the stream-lines, and the radii $\theta = $ const. as the equipotential lines. In both cases the velocity at a distance r from the origin is $\dfrac{\mu}{r}$; we must therefore suppose the origin excluded (*e.g.* by drawing a small circle round it) from the region occupied by the fluid. In the second case, already discussed in Art. 34, the motion is cyclic; the circulation in any circuit surrounding the origin being $2\pi\mu$.

82. *Example* 3. Let us write

$$\phi + i\psi = \mu \log \frac{x + iy - a}{x + iy + a}.$$

If r_1, r_2 denote the distances of any point in the plane xy from the points $(\pm a, 0)$, and θ_1, θ_2 the angles which these distances make with the positive direction of x, we have

$$\frac{x + iy - a}{x + iy + a} = \frac{r_1 e^{i\theta_1}}{r_2 e^{i\theta_2}},$$

and therefore

$$\phi = \mu \log \frac{r_1}{r_2}, \quad \psi = \mu (\theta_1 - \theta_2).$$

The curves $\psi = $ const., *i.e.* $\theta_1 - \theta_2 = $ const., are circles passing through the points $(\pm a, 0)$; the curves $\phi = $ const. are the system of circles orthogonal to these. Either of these systems of circles may be taken as the equipotential curves, and the other system will then compose the stream-lines. In either case the velocity at the points $(\pm a, 0)$ will be infinite; so that we must exclude these points (*e.g.* by small closed curves drawn round them) from the region to which the formulæ apply, which thus becomes triply-connected. If the curves $\dfrac{r_1}{r_2} = $ const. be taken as the stream-lines, the circulation in any circuit embracing the first only of the above points is $2\pi\mu$; that in one embracing the second point only is $-2\pi\mu$; whilst that in a circuit embracing both is zero.

83. *Example* 4. Assume

$$\phi + i\psi = \mu \log (x + iy - a)(x + iy + a).$$

If r_1, r_2, θ_1, θ_2 have the same meanings as in the last example, this gives

$$\phi = \mu r_1 r_2, \quad \psi = \mu (\theta_1 + \theta_2) \ldots\ldots\ldots\ldots(28).$$

The curves $r_1 r_2 = $ const. are a system of lemniscates whose poles are at the points $(\pm a, 0)$. The curves $\theta_1 + \theta_2 = $ const., *i.e.*

$$\text{arc tan} \frac{y}{x - a} + \text{arc tan} \frac{y}{x + a} = \text{const.},$$

or
$$\arctan \frac{2xy}{x^2 - y^2 - a^2} = \text{const.,}$$

are a system of rectangular hyperbolas, orthogonal to the above system of lemniscates, and passing through their poles. A drawing of both systems of curves is given by Lamé*.

The formulæ (28) make the velocity infinite at the poles, which must therefore be excluded from the region to which the formulæ apply. If the lemniscates be taken as the stream-lines, the velocity-potential $\mu (\theta_1 + \theta_2)$ is a cyclic function; the circulation in any circuit embracing one pole only is $2\pi\mu$, that in a circuit embracing both poles is $4\pi\mu$.

84. *Example 5.* Assume

$$\frac{dw}{dz} = u \log \frac{z - a}{z + a} \quad\dots\dots\dots\dots\dots\therefore (29),$$

$$= \mu \log \frac{r_1 e^{i\theta_1}}{r_2 e^{i\theta_2}},$$

in the notation of Art. 82. Since

$$\frac{dw}{dz} = \frac{d\phi}{dx} + i \frac{d\psi}{dx} = u - iv,$$

this gives

$$u = \mu \log \frac{r_1}{r_2}, \quad v = \mu (\theta_2 - \theta_1) \quad\dots\dots\dots\dots (30).$$

If we suppose the region occupied by the fluid to have the axis of x as a boundary, there will be no ambiguity in these values of u, v. Moreover, u, v will be everywhere finite and continuous except on the axis of x. When $y = 0$, and $x > a$, then $\theta_1 = \theta_2 = 0$; and when $y = 0$, $x < -a$, then $\theta_1 = \theta_2 = \pi$. In each case $v = 0$. When $y = 0$, and $a > x > -a$, we have $\theta_1 = \pi$, $\theta_2 = 0$, and therefore $v = -\mu\pi$. We have thus the solution of the following problem: An infinite mass of liquid bounded by an infinite rigid plane but otherwise unlimited is initially at rest, and a strip of this plane of breadth $2a$ is supposed detached from the remainder and suddenly pushed inwards with velocity $\mu\pi$; to find the motion produced in the fluid. In the above formulæ the rigid plane corresponds to the axis of x, and the fluid lies to the negative side of the latter.

It appears from (30) that at the edges of the strip u is infinite.

---

* *Leçons sur les Coordonnées curvilignes*, p. 223.

This example is merely given as an instance of discontinuous boundary-conditions.

The values of ϕ, ψ can be at once found, if required, by integrating (30).

85. A very general formula for the functions ϕ, ψ may be obtained as follows. It may be shewn that if a function $f(z)$ be finite, continuous, and single-valued, and have its first derivative finite, at all points of a space included between two concentric circles about the origin, its value at any point of the space can be expanded in the convergent form

$$f(z) = A_0 + A_1 z + A_2 z^2 + \dots$$
$$+ B_1 z^{-1} + B_2 z^{-2} + \dots \dots \dots \dots (31).$$

If the above conditions be satisfied at all points within a circle having the origin as centre, we retain only the ascending series; if at all points without such a circle, the descending series, with the addition of the constant A_0, is sufficient. If the conditions be fulfilled for all points of the plane xy without exception, $f(z)$ can be no other than a constant A_0.

Putting $f(z) = \phi + i\psi$, introducing polar co-ordinates, and writing the complex constants A_n, B_n, in the forms $P_n + iQ_n$, $R_n + iS_n$, respectively, we obtain

$$\phi = P_0 + \Sigma_1^\infty r^n (P_n \cos n\theta - Q_n \sin n\theta) + \Sigma_1^\infty r^{-n} (R_n \cos n\theta + S_n \sin n\theta)\}$$
$$\psi = Q_0 + \Sigma_1^\infty r^n (Q_n \cos n\theta + P_n \sin n\theta) + \Sigma_1^\infty r^{-n} (S_n \cos n\theta - R_n \sin n\theta)\}$$
$$\dots \dots \dots \dots (32).$$

These formulæ are convenient in treating problems where we have the value of ϕ, or of $\dfrac{d\phi}{dn}$, given over the circular boundaries. This value may be expanded for each boundary in a series of sines and cosines of multiples of θ, by Fourier's theorem. The series thus found must be equivalent to those obtained from (32), whence, equating separately coefficients of $\sin n\theta$ and $\cos n\theta$, we obtain four systems of linear equations to determine

$$P_n, \; Q_n, \; R_n, \; S_n.$$

86. *Example* 6. An infinitely long circular cylinder of radius

a is moving with velocity V perpendicular to its length, in an infinite mass of liquid which is at rest at infinity; to find the motion of the fluid supposing it to have been started from rest.

The motion will evidently be in two dimensions. Let the origin be taken in the axis of the cylinder, and the axes of x, y in a plane perpendicular to its length. Further let the axis of x be in the direction of the velocity V. The motion having originated from rest will necessarily be irrotational, and ϕ will be single-valued. Also, since $\int \dfrac{d\phi}{dn} ds$, taken round the section of the cylinder is zero, ψ is also single-valued (see Art. 69), so that the formulæ (32) apply. Moreover, since $\dfrac{d\phi}{dn}$ is given at every point of the cylinder, viz.

$$\frac{d\phi}{dr} = V \cos \theta, \text{ when } r = a \ldots\ldots\ldots\ldots (33),$$

the problem is determinate, by Art. 71. Since the region occupied by the fluid extends to infinity we must in (32) omit the coefficients P_n, Q_n. The condition (33) then gives

$$V \cos \theta = - \Sigma_1^\infty n a^{-n-1} (R_n \cos n\theta + S_n \sin n\theta);$$

which can be satisfied only by making $R_1 = - Va^2$, and all the other coefficients zero. The complete solution is therefore given by

$$\phi = - \frac{Va^2}{r} \cos \theta, \quad \psi = \frac{Va^2}{r} \sin \theta \ldots\ldots\ldots\ldots (34).$$

These formulæ coincide with those of Art. 80 (e).

As this case is one which is readily comparable with experiment, we will calculate the effect of the pressure of the fluid on the surface of the cylinder. The formula (4) of Art. 25 gives

$$\frac{p}{\rho} = C - \frac{d\phi}{dt} - \tfrac{1}{2} q^2 \ldots\ldots\ldots\ldots\ldots (35),$$

where we have omitted the term due to the external impressed forces, the effect of which can be calculated by the ordinary rules of Hydrostatics. The term $\dfrac{d\phi}{dt}$ in (35) expresses the rate at which ϕ is increasing at a fixed point of space, whereas the value of ϕ

in (34) is referred to an origin which is in motion with the velocity V parallel to x. Hence

$$\frac{d\phi}{dt} = -\frac{dV}{dt} \cdot \frac{a^2}{r} \cos\theta - V\frac{d\phi}{dx},$$

where

$$\frac{d\phi}{dx} = \frac{d\phi}{dr}\cos\theta - \frac{d\phi}{r d\theta}\sin\theta = \frac{Va^2}{r^2}\cos 2\theta.$$

Also

$$q^2 = V^2 \frac{a^4}{r^4}.$$

The pressure at any point of the cylindrical surface is therefore

$$p = \rho\left(C + \frac{dV}{dt}\cos\theta + V^2\cos 2\theta - \tfrac{1}{2}V^2\right)\ldots\ldots\ldots(36).$$

The resultant pressure on a length of the cylinder is evidently parallel to x; to find its amount per unit length we must multiply (36) by $-ad\theta . \cos\theta$ and integrate with respect to θ between the limits 0 and 2π. The only term which gives a result different from zero is the second, which gives

$$-\pi\rho a^2 \frac{dV}{dt},$$

or

$$-M'\frac{dV}{dt}\ldots\ldots\ldots\ldots\ldots\ldots\ldots\ldots(37),$$

if M' be the mass in unit length of the fluid displaced by the cylinder. Compare Art. 105.

If in the above example we impress on the fluid and the cylinder a velocity $-V$ in the direction of x, we have the case of a current flowing with velocity V past a fixed cylindrical obstacle. Adding to ϕ and ψ the terms $-Vx$ and $-Vy$, respectively, we get

$$\phi = -V\left(r + \frac{a^2}{r}\right)\cos\theta, \quad \psi = -V\left(r - \frac{a^2}{r}\right)\sin\theta.$$

If no external forces act, and if V be constant, we find for the resultant pressure on the cylinder the value zero.

87. To render the formula (31) capable of representing *any* case of irrotational motion in the space between two concentric circles, we must add to the right-hand side the term

$$A \log z \ldots\ldots\ldots\ldots\ldots\ldots\ldots\ldots(38).$$

If $A = P + iQ$, the corresponding terms in ϕ, ψ are

$$P \log r - Q\theta, \quad P\theta + Q \log r,$$

respectively. The meaning of these terms will appear from Example 2 above. $2\pi P$ is the cyclic constant of ψ, *i.e.* (Art. 69) the total flux across the inner (or outer) circle; and $-2\pi Q$, the cyclic constant of ϕ, is the circulation in any circuit embracing the origin.

The formula (31), as amended by the addition of the term (38), may readily be generalized so as to apply to any case of irrotational motion in a region with circular boundaries, one of which encloses all the rest. In fact, corresponding to each internal boundary we have a series of the form

$$A \log (z - c) + \frac{A_1}{z - c} + \frac{A_2}{(z - c)^2} + \&\text{c.},$$

where c, $= a + ib$ say, refers to the centre, and the coefficients A, A_1, A_2, &c. are in general complex quantities. The difficulty however of determining these coefficients so as to satisfy given boundary conditions is now so great as to render this method of very little utility.

Indeed the determination of the irrotational motion of a liquid subject to given boundary conditions is a problem whose exact solution can be effected by direct processes in only a very few cases. Most of the cases for which we know the solution have been obtained by an inverse process; viz. instead of trying to find a solution of the equation (5a) or (7) satisfying given boundary conditions, we take some known solution of the differential equations and enquire what boundary conditions it can be made to satisfy. In this way we may obtain some interesting results in the following two important cases of the general problems in two dimensions.

88. *Case* I. The boundary of the fluid consists of a rigid cylindrical surface which is in motion with velocity V in a direction perpendicular to its length.

Let us take as axis of x the direction of this velocity V, and let ds be an element of the section of the surface by the plane xy.

Then at all points of this section

$$\frac{d\psi}{ds} = \text{velocity of the fluid in the direction of the normal,}$$

$$= \text{velocity of the boundary normal to itself,}$$

$$= V \frac{dy}{ds}.$$

Integrating along the section, we have

$$\psi = Vy + \text{const.} \dots\dots\dots\dots\dots\dots(39).$$

If we take any possible form of ψ, the equation (39) is the equation of a system of curves each of which would by its motion parallel to x produce the set of stream-lines defined by $\psi = \text{const.}$ We give a few examples.

(a) If we choose for ψ the form $Vy + \text{const.}$, then (39) is satisfied identically for all forms of the boundary. Hence the fluid contained within a cylinder of any shape which has a motion of translation only may move as a solid body. If, further, the cylindrical space occupied by the fluid be simply-connected, this is the only kind of motion possible. This is otherwise evident from Art. 49; for the motion of the fluid and the solid as one mass evidently satisfies the boundary conditions, and is therefore the only solution which the problem admits of.

(b) Let $\psi = \dfrac{A}{r} \sin \theta$ (Example 1); then (39) becomes

$$\frac{A}{r} \sin \theta = Vr \sin \theta + \text{const.}$$

In this system of curves is included a circle of radius a, provided

$$\frac{A}{a} = Va.$$

Hence the motion produced in an infinite mass of liquid by a circular cylinder moving through it with velocity V perpendicular to its length, is given by

$$\psi = \frac{Va^2}{r} \sin \theta,$$

which agrees with (34).

(c) With the same notation as in Example 3 let us assume

$$\psi = A (\theta_1 - \theta_2).$$

The equation of the curves by whose motion parallel to x this system of stream-lines could be produced, is by (39)

$$y = k\,(\theta_1 - \theta_2) + C \dotfill (40),$$

where k is written for $\dfrac{A}{V}$.

If we put $C = 0$ we obtain an oval curve symmetrical with respect to x and y. The curve corresponding to any other value of C is symmetrical with respect to the axis of y, and has the line $y = C$ as an asymptote towards $x = \pm \infty$. The curves for which C is the same in magnitude, but of opposite sign, are symmetrically situated on opposite sides of the axis of x. The points $(\pm a, 0)$ which in Example 3 are taken as origins of θ_1, θ_2, may be called the foci of the above system of curves. By varying the constants k and a the forms of the curves can be varied indefinitely. From their resemblance (within certain limits as to the relative magnitudes of k and a) to the lines of ships, they have been called 'bifocal neoïds,' by Prof. Rankine*, who investigated their properties with a view to obtaining theoretical guidance as to what proportions are to be observed in designing a ship in order to reduce as much as possible the resistances due to waves, surface-friction, &c.

(d)   Let ξ, η be two new variables connected with x, y by the relation

$$x + iy = c \sin(\xi + i\eta).$$

This gives

$$\left.\begin{aligned}
x &= c \sin \xi \cdot \frac{e^{\eta} + e^{-\eta}}{2}, \\[2mm]
y &= c \cos \xi \cdot \frac{e^{\eta} - e^{-\eta}}{2}.
\end{aligned}\right\} \dotfill (41).$$

Eliminating ξ, we have

$$\frac{x^2}{c^2 \left(\dfrac{e^{\eta} + e^{-\eta}}{2}\right)^2} + \frac{y^2}{c^2 \left(\dfrac{e^{\eta} - e^{-\eta}}{2}\right)^2} = 1 \dotfill (42),$$

and eliminating η,

$$\frac{x^2}{c^2 \sin^2 \xi} - \frac{y^2}{c^2 \cos^2 \xi} = 1.$$

Hence the curves $\xi = \text{const.}$, $\eta = \text{const.}$ are confocal hyperbolas and

* *Phil. Trans.*, 1864.

ellipses, respectively; the distance between the common foci
being $2c$.

Now let us assume

$$\phi + i\psi = Cie^{i(\xi + i\eta)},$$

where C is some real constant. This makes $\phi + i\psi$ a function of
$x + iy$; and the value of ψ is

$$\psi = Ce^{-\eta}\cos \xi,$$

so that (39) becomes

$$Ce^{-\eta}\cos \xi = Vc \cos \xi . \frac{e^{\eta} - e^{-\eta}}{2} + \text{const.}$$

In this system of curves is included the ellipse whose parameter η
is determined by

$$Ce^{-\eta} = Vc . \frac{e^{\eta} - e^{-\eta}}{2} .$$

If a, b be the semi-axes of this ellipse, we have, by (42),

$$a = c . \frac{e^{\eta} + e^{-\eta}}{2} , \qquad b = c . \frac{e^{\eta} - e^{-\eta}}{2} ,$$

so that

$$C = \frac{Vbc}{a - b} = Vb \sqrt{\left(\frac{a + b}{a - b}\right)} .$$

Hence the formula

$$\psi = Vb \sqrt{\left(\frac{a + b}{a - b}\right)} e^{-\eta}\cos \xi \ldots\ldots\ldots\ldots (43)$$

gives the motion of an infinite mass of liquid produced by an
elliptic cylinder whose semi-axes are a, b, moving parallel to its
major axis with velocity V.

That the above formula makes the velocity zero at infinity
appears from the consideration that when η is large, dx and dy are
of the same order as $e^{\eta}d\eta$ or $e^{\eta}d\xi$, so that $\frac{d\psi}{dx}$, $\frac{d\psi}{dy}$ are of the

order $e^{-2\eta}$, or $\frac{1}{r^2}$, ultimately, where r denotes the distance of any
point from the axis of the cylinder.

If the motion of the cylinder were parallel to its minor axis,
the formula would be

$$\psi = - Va \sqrt{\left(\frac{a + b}{a - b}\right)} e^{-\eta}\sin \xi \ldots\ldots\ldots\ldots (44).$$

Drawings of the curves $\phi = \text{const.}$, $\psi = \text{const.}$, in the above cases are given in the *Quarterly Journal of Mathematics*, December, 1875. These curves are the same for all confocal ellipses; so that the formulæ hold even when the generating ellipse reduces to a straight line joining the foci. In this case (44) becomes

$$\psi = - Vce^{-\eta} \sin \xi \quad\text{.....................} (45),$$

which would, except for the reasons stated in Art. 30, give the motion produced in an infinite mass of liquid by an infinitely long lamina of breadth $2c$ moving perpendicular to itself with velocity V. Since however (45) makes the velocity at the edges of the lamina infinite, this solution is destitute of practical value.

When $c = 0$ the problem of this section becomes that of (*b*) above. The student may, as an exercise, work out the transformation of (43) into (34).

89.   *Case* II.   The boundary of the fluid consists of a rigid cylindrical surface rotating with angular velocity ω about an axis parallel to its length.

Taking the origin in the axis of rotation, and the axes of x, y in a perpendicular plane, we have, with the same notation as before,

$$\frac{d\psi}{ds} = \text{velocity in the direction of the normal to the boundary}$$

$$= - \omega y \frac{dy}{ds} - \omega x \frac{dx}{ds},$$

$\Big($the component velocities of a point of the boundary being

$- \omega y$, ωx, and the direction-cosines of the normal $\dfrac{dy}{ds}$, $-\dfrac{dx}{ds}\Big)$.

Integrating we have, at all points of the boundary,

$$\psi = - \tfrac{1}{2}\omega (x^2 + y^2) + \text{const.} \quad\text{...............} (46).$$

If we assume any possible form of ψ, this will give us the equation of a series of curves, each of which would by rotation round the origin, produce the system of stream-lines determined by ψ.

As examples we may take the following :

(a)   If we assume

$$\psi = Ar^3 \cos 2\theta = A(x^2 - y^2),$$

the equation (46) then becomes

$$(\tfrac{1}{2}\omega + A)\, x^2 + (\tfrac{1}{2}\omega - A)\, y^2 = C,$$

which, for any given value of A represents a system of similar and coaxial conic sections.   That this system may include the ellipse

$$\frac{x^2}{a^2} + \frac{y^2}{b^2} = 1,$$

we must have

$$(\tfrac{1}{2}\omega + A)\, a^2 = (\tfrac{1}{2}\omega - A)\, b^2,$$

or

$$A = -\tfrac{1}{2}\omega \cdot \frac{a^2 - b^2}{a^2 + b^2}.$$

Hence

$$\psi = -\tfrac{1}{2}\omega \cdot \frac{a^2 - b^2}{a^2 + b^2}(x^2 - y^2)$$

gives the motion of a liquid contained within a hollow elliptic cylinder whose semi-axes are a, b, produced by the rotation of the cylinder about its axis with angular velocity ω.

The corresponding formula for ϕ is

$$\phi = \omega \cdot \frac{a^2 - b^2}{a^2 + b^2} \cdot xy.$$

The angular momentum of the fluid, per unit length of the cylinder, about the axis of rotation, is

$$\rho \iint \left(x \frac{d\phi}{dy} - y \frac{d\phi}{dx} \right) dx\, dy = \tfrac{1}{4}\rho\omega \cdot \frac{a^2 - b^2}{a^2 + b^2} \cdot \pi ab\, (a^2 - b^2).$$

Hence the cylinder rotates under the action of any external forces exactly as if the fluid were replaced by a solid whose moment of inertia about the axis of rotation is

$$\tfrac{1}{4}M \frac{(a^2 - b^2)^2}{a^2 + b^2}$$

per unit length, M being the mass per unit length of the fluid.

(b)   Let us assume

$$\psi = Ar^3 \cos 3\theta = A(x^3 - 3xy^2).$$

The equation of the boundary (46) then becomes

$$A(x^3 - 3xy^2) + \tfrac{1}{2}\omega(x^2 + y^2) = C \ldots\ldots\ldots\ldots (47).$$

Let us choose the constants so that the straight line $x = a$ may be part of the boundary. The conditions for this are

$$Aa^3 + \tfrac{1}{2}\omega a^2 = C, \quad -3aA + \tfrac{1}{2}\omega = 0.$$

Substituting the values of A, C hence derived in (47), we have

$$x^3 - a^3 - 3xy^2 + 3a\,(x^2 - a^2 + y^2) = 0.$$

Dividing out by $x - a$, we get

$$x^2 + 4ax + 4a^2 = 3y^2,$$

or

$$x + 2a = \pm \sqrt{3} \cdot y.$$

The rest of the boundary consists then of two straight lines passing through the point $(-2a, 0)$, and inclined at angles of 30° to the axis of x. The complete boundary, therefore, is composed of three straight lines forming an equilateral triangle, the origin being at the centre of gravity.

We have thus obtained the formulæ for the motion of the fluid contained within a vessel in the form of an equilateral prism, when the latter is set in motion with angular velocity ω about an axis parallel to its length and passing through the centre of gravity of its section; viz. we have

$$\psi = \tfrac{1}{6} \frac{\omega}{a} r^3 \cos 3\theta, \quad \phi = -\tfrac{1}{6} \frac{\omega}{a} r^3 \sin 3\theta,$$

where $2\sqrt{3}a$ is the length of a side of the prism.

The problem of fluid motion in a rotating cylindrical case is to a certain extent mathematically identical with that of the torsion of a uniform rod or bar[*]. The above cases (a), (b) are mere adaptations of two of M. de Saint-Venant's solutions of the latter problem.

(c) With the same notation as in Art. 88 (d), let us assume

$$\phi + i\psi = Ce^{2i(\xi + i\eta)}.$$

We have, from (41),

$$x^2 + y^2 = \tfrac{1}{4}c^2\,(e^{2\eta} - 2\cos 2\xi + e^{-2\eta}).$$

Hence (46) becomes

$$Ce^{-2\eta}\cos 2\xi + \tfrac{1}{8}\omega c^2\,(e^{2\eta} - 2\cos 2\xi + e^{-2\eta}) = \text{const.}$$

---

[*] See Thomson and Tait, *Natural Philosophy*, Art. 704, *et seq.*

This system of curves includes the ellipse whose parameter is η provided

$$Ce^{-2\eta} = \tfrac{1}{4}\omega c^2,$$

or, using the values of a, b already given,

$$C = \tfrac{1}{4}\omega\,(a+b)^2,$$

so that

$$\psi = \quad \tfrac{1}{4}\omega\,(a+b)^2\,e^{-2\eta}\cos 2\xi,$$

$$\phi = -\tfrac{1}{4}\omega\,(a+b)^2\,e^{-2\eta}\sin 2\xi.$$

By the same reasoning as in the last article we see that at a great distance from the origin the velocity is of the order $\dfrac{1}{r^3}$.

The above formulæ therefore give the motion of an infinite mass of liquid, otherwise at rest, produced by the rotation of an elliptic cylinder about its axis with angular velocity ω.

A drawing of the stream-lines in this case is given in the *Quarterly Journal of Mathematics*, December, 1875.

90. If w be a function of z, it follows at once from the definition of Art. 74 that z is a function of w. The latter form of the assumption is sometimes more convenient analytically than the former.

The relations (11) are then replaced by

$$\frac{dx}{d\phi} = \frac{dy}{d\psi}, \quad \frac{dx}{d\psi} = -\frac{dy}{d\phi} \quad\ldots\ldots\ldots\ldots (48).$$

Also, since

$$\frac{dw}{dz} = \frac{d\phi}{dx} + i\frac{d\psi}{dx} = u - iv,$$

we have

$$\frac{dz}{dw} = \frac{1}{u - iv} = \frac{1}{q}\left(\frac{u}{q} + i\frac{v}{q}\right),$$

where q is the resultant velocity at (x, y). Hence if the properties of the function $\dfrac{dz}{dw}$ ($= \zeta$, say,) be exhibited graphically in the manner already explained, the vector drawn from the origin to any point in the plane of ζ will agree in direction with, and be in magnitude the reciprocal of, the velocity at the corresponding point of the plane of z.

Again since $\dfrac{1}{q}$ is the modulus of $\dfrac{dz}{dw}$, *i.e.* of $\dfrac{dx}{d\phi} + i\dfrac{dy}{d\phi}$, we have

$$\frac{1}{q^2} = \left(\frac{dx}{d\phi}\right)^2 + \left(\frac{dy}{d\phi}\right)^2 \dots\dots\dots\dots\dots(49),$$

which may, by (48), be put into the equivalent forms

$$\frac{1}{q^2} = \left(\frac{dx}{d\phi}\right)^2 + \left(\frac{dx}{d\psi}\right)^2 = \left(\frac{dy}{d\phi}\right)^2 + \left(\frac{dy}{d\psi}\right)^2$$
$$= \left(\frac{dx}{d\psi}\right)^2 + \left(\frac{dy}{d\psi}\right)^2 = \frac{dx}{d\phi}\frac{dy}{d\psi} - \frac{dx}{d\psi}\frac{dy}{d\phi} \dots\dots(50).$$

The last formula, viz.

$$\frac{1}{q^2} = \frac{d(x,\ y)}{d(\phi,\ \psi)},$$

simply expresses the fact that corresponding elementary areas in the planes of z and w are in the ratio of the square of the modulus of $\dfrac{dz}{dw}$ to unity.   Compare Art. 76.

91. *Example* 7.   Assume

$$z = c \sin w,$$

or

$$x = c \cdot \frac{e^\psi + e^{-\psi}}{2} \sin \phi,$$

$$y = c \cdot \frac{e^\psi - e^{-\psi}}{2} \cos \phi.$$

The curves $\psi = $ const. are, Art. 88 (*d*), a system of confocal ellipses, and the curves $\phi = $ const. a system of confocal hyperbolas; the common foci of the two systems being the points $(\pm c, 0)$.

Since at the foci we have $\phi = \frac{1}{2}(2n+1)\pi$, $\psi = 0$, n being some integer, we see by (49) that the velocity is infinite there. We must therefore exclude these points from the region to which our formulæ apply. If the ellipses be taken as the stream-lines the motion is cyclic; the circulation in any circuit embracing either focus alone is $-\pi$, that in a circuit embracing both is -2π.

At an infinite distance from the origin ψ is infinite, and the velocity zero.

When $c = 0$ this case coincides with that of Example 2.   We leave it as an exercise for the student to deduce the formulæ of that example from those of the present article.

If we take the hyperbolas as the stream-lines, the portions of the axis of x which lie beyond the points $(\pm c, 0)$ may be taken as fixed boundaries. We obtain in this manner the case of a liquid flowing from one side to the other of a rigid plane, through an aperture of breadth $2c$ made in the plane; but since the velocity at the edges of the aperture is infinite, this kind of motion cannot be realized with actual fluids.

92. *Example* 8. Let

$$z = A (w + e^w),$$

whence

$$\left. \begin{array}{l} x = A\phi + Ae^\phi \cos \psi \\ y = A\psi + Ae^\phi \sin \psi \end{array} \right\} \quad \ldots\ldots\ldots\ldots\ldots\ldots (51).$$

Along the stream-lines $\psi = \pm A\pi$, we have

$$x = A (\phi - e^\phi), \quad y = \pm A\pi.$$

As ϕ increases from $-\infty$ through zero to $+\infty$, x increases from $-\infty$, reaches a certain maximum value, and then goes back to $-\infty$. The maximum value is readily found to be when $\phi = 0$, and is $-A$. Hence the portions of the straight lines $y = \pm A\pi$ which lie between $x = -\infty$, and $x = -A$, may be taken as fixed boundaries. Let us next trace the course of a stream-line infinitely near to one of the former; say $\psi = A (\pi - \alpha)$, where α is infinitesimal. This gives

$$x = A (\phi + e^\phi \cos \psi), \quad y = A\pi - A\alpha + A\alpha e^\phi,$$

approximately. As ϕ increases from $-\infty$, x increases, whilst y remains at first approximately constant and equal to $A (\pi - \alpha)$; when, however x approaches its maximum value, y increases to the value $A\pi$. As ϕ increases beyond the value zero, x diminishes, whilst the excess of y over $A\pi$ slowly but continuously increases.

The formulæ (51) express then the motion of a liquid flowing from a canal bounded by two parallel planes into open space*. We see, however, from (50), that the velocity at the edges of these planes (where $\phi = 0$, $\psi = \pm \pi$) is infinite; so that the motion

---

* The above example is due to Helmholtz, *Phil. Mag.* Nov. 1868. A drawing of the curves $\phi = \text{const.}$, $\psi = \text{const.}$, is given in Maxwell's *Electricity and Magnetism*, Vol. I., Plate XII.

cannot be realized, for the reasons explained in Art. 30. If however the motion be very slow, we may take two stream-lines very near to $\psi = \pm \pi$ as fixed boundaries, and so obtain a possible case.

93. *Example* 9. Kirchhoff has, *à propos* of certain problems to be discussed below, given a method by which the determination of the motion in several cases of interest may be readily effected. The method rests upon the property of the function ζ explained in Art. 90. If the boundaries of the fluid be fixed and rectilinear, the corresponding lines in the plane of ζ, which are also straight, are easily laid down. Also, since the fixed boundaries are stream-lines, the corresponding lines in the plane of w are straight lines $\psi = \text{const.}$ It is then in many cases not difficult to frame an assumption of the form

$$\zeta = f(w),$$

by which the correspondence of these lines in the planes of ζ and w may be established. The relation between z and w is then to be found by integration.

Example 8, above, is very easily treated in this manner. We take however a somewhat less simple case; viz. that in which a current flows from a uniform canal into an open space which is bounded by an infinite plane perpendicular to the length of the canal, and in which the mouth of the latter lies. See Fig. 6.

Fig. 6.

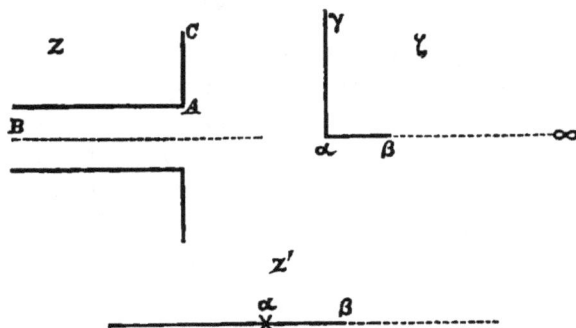

The middle line of the canal is evidently a stream-line; say that for which $\psi = 0$. Also for the stream-line BAC let $\psi = \pi$;

on account of the symmetry we need only consider the motion between these two lines. If the velocity in the canal at a distance from the mouth be taken to be unity, the half-breadth of the canal will be π. The boundaries in the plane of ζ are shewn in the figure, corresponding points in the planes of z and ζ being indicated by corresponding Roman and Greek letters. The point A corresponds to the origin of ζ because the velocity there is infinite. (Art. 80.)

We have now to connect ζ and w by a relation such that $\gamma\alpha\beta$ and $\beta\infty$ shall correspond to the two straight lines $\psi = 0$, $\psi = \pi$. If we assume $z' = \zeta^2$, then in the plane of z' the lines $\gamma\alpha\beta$ and $\beta\infty$ become parts of the same straight line. The assumption $z' = 1 + e^w$ then converts these two parts into the straight lines $\psi = 0$, $\psi = \pi$. See Art. 78. We have then

$$\frac{dz}{dw} = \zeta = \sqrt{1 + e^w},$$

whence

$$z = 2\sqrt{1 + e^w} + \log \frac{\sqrt{1 + e^w} - 1}{\sqrt{1 + e^w} + 1}.$$

The constant of integration is so chosen that the origin of z (hitherto arbitrary) shall be at the intersection of the middle line of the canal with the plane of its mouth.

Discontinuous Motions*.

94. We have had frequent occasion to remark, concerning forms of fluid motion which we have obtained, that they cannot be realized in practice on account of the infinite velocity and consequent negative pressure which they would involve at some point of the boundary. We are led to solutions of this nugatory character whenever a sharp projecting edge forms part of the boundary. Edges of absolute geometrical sharpness do not of course occur in practice; but even if the edge be slightly rounded, (as for instance in Example 8 above, by the substitution of a neighbouring stream-line as the fixed boundary,) the velocity in the immediate neighbourhood will, unless the motion be every-

---

* Helmholtz, *l.c.* Art. 92.

where else exceedingly slow, be very great, and so will still transgress the limit pointed out in Art. 30.

It is a matter of ordinary observation that under such circumstances a surface of discontinuity, beginning at the sharp edge, is formed. Thus the current issuing from an orifice in the thin wall of a vessel does not, as would appear from Example 7, spread out and follow the walls, but forms a compact stream bounded on all sides by fluid sensibly at rest. Similarly, the current issuing from a straight pipe or canal does not spread out in all directions in the manner indicated in Example 9, but forms, at all events for a short distance, a uniform stream whose boundaries are prolongations of the sides of the canal. As practical exemplifications of these statements we may point to the smoke issuing from a chimney, and to the motion of a rapid torrent through a bridge whose span is considerably less than the breadth of the channel below.

It is not very easy to form, from theory, a precise idea of the manner in which the existence of these surfaces of discontinuity is brought about. If the motion in any of the cases above referred to be generated gradually from rest, as for instance in the case of Example 9 by the motion of a piston fitting the canal, then if the edges be slightly rounded the continuous motion already discussed will in the first instance be possible. As however the motion of the piston is accelerated, a time arrives when the pressure at the edge sinks to zero. The boundary-conditions are then altered and the nature of the analytical problem is entirely changed. Helmholtz supposes that at the points of zero pressure the values of the derivatives $\dfrac{dp}{dx}$, $\dfrac{dp}{dy}$, $\dfrac{dp}{dz}$, become discontinuous, and that it is to these discontinuous components of the total force acting on a fluid element that the generation of the discontinuous motion which is actually produced is to be ascribed.

The conditions to be satisfied at a surface of discontinuity are easily found. We have of course the kinematical relation (13) of Art. 10; and the dynamical condition is that the pressure at every point of the surface must be the same on both sides. If the motion be steady, this requires that the values of the squares of the velocities on the two sides should differ by a constant quantity. If further, as occurs in most cases of practical interest, the fluid on one side of the surface be sensibly at rest, the conditions reduce to these, that the velocity must be wholly tangential to the surface, and of constant magnitude.

Kirchhoff's Method.

95. Kirchhoff has applied the method already partially explained to obtain forms of discontinuous motion (in two dimensions) satisfying the conditions just stated. The fluid is supposed bounded by two stream-lines $\psi = \alpha$, $\psi = \beta$, consisting partly of fixed boundaries and partly of lines of constant (say unit) velocity. The lines for which $q = 1$ are represented in the plane of ζ by arcs of a circle of unit radius having the origin as centre; whilst the fixed boundaries, if straight (as we shall suppose them to be), become radii of this circle. The points where the radii meet the circle correspond to the points where the limiting stream-lines change their character. We have then to frame an assumption of the form

$$\zeta = f(w)\dots\dots\dots\dots\dots\dots(52),$$

such that the portion of the plane of ζ external to the above circle and included between the two radii shall correspond to the portion of the plane of w included between the two parallel straight lines $\psi = \alpha$, $\psi = \beta$. It is further necessary (see Art. 79), that the function $f(w)$ shall have no branch-points within the portion of the plane of w considered, although such points may occur on the boundary of this portion.

It is found that this problem may be reduced to a particular case of the following:—To connect two complex variables z, z' by a functional relation such that any given lune in the plane of z shall correspond to any given lune in the plane of z', and any three points on the perimeter of the one lune to any three points on that of the other. By a 'lune' is here meant the closed figure formed by two circular arcs which meet but do not cross. By the 'angle of a lune' we shall understand the angle contained by the arcs at either intersection.

To solve the above problem we remark in the first place that any assumption of the form

$$Z' = \frac{AZ + B}{CZ + D}, \text{ or } Z = \frac{-DZ' + B}{CZ' - A} \dots\dots\dots(53),$$

transforms circles into circles. For suppose the point Z' to describe a circle about any point C' as centre. We have then

$$\text{mod } (Z' - C') = \text{const.,}$$

and therefore, if by a change of constants $Z' - C'$ be put into the form

$$K \frac{Z - C_1}{Z - C_2},$$

$$\text{mod } (Z - C_1) : \text{mod } (Z - C_2) = \text{const.}$$

Hence the point Z moves so that its distances from the two points C_1, C_2 are in a constant ratio; *i.e.*, by a well-known theorem of elementary geometry, it describes a circle. To a lune in the plane of Z corresponds then a lune of the same angle in the plane of Z'. The three ratios $A : B : C : D$ may be so chosen as to make any three points in the one plane correspond to any three points in the other, when the circles determined by these triads will also correspond. Hence to establish a correspondence between any two lunes *of the same angle* we have only to determine the above ratios so that the angular points of the one shall correspond to the angular points of the other, and any third point on the perimeter of the one to any third point on that of the other.

As a particular case, the assumption

$$\mathfrak{Z} = \frac{z - c_1}{z - c_2}$$

transforms any lune whose angular points are at c_1, c_2 into a lune in the plane of $\mathfrak{Z}$ having its angular points at 0 and ∞, *i.e.* into two straight lines radiating from the origin, and making an angle equal to that of the lune.

If we now assume

$$Z = \mathfrak{Z}^n,$$

these straight lines become, in the plane of Z, straight lines inclined at an angle n times as great. Compare Art. 80. If we then make $n = \dfrac{\pi}{\alpha}$, so that

$$Z = \left(\frac{z - c_1}{z - c_2}\right)^{\frac{\pi}{\alpha}} \quad \text{.........................(54)},$$

a lune of angle α having its angular points at c_1, c_2 becomes in the plane of Z one straight line, and therefore by (53) in the plane of Z' a circle. The constants in (53) may be so determined that any three points on the perimeter of the lune correspond to any three points in the plane of Z'.

Since, in the same way, the assumption

$$Z' = \left(\frac{z' - c_1'}{z' - c_2'}\right)^{\frac{\pi}{\alpha'}} \quad \text{.........................(55)}$$

transforms a lune of angle α' having its extremities at c_1', c_2' into a straight line in the plane of Z', we see that (53), with (54) and (55), transforms a lune of angle α having its extremities at c_1, c_2 into a lune of angle α' having its extremities at c_1', c_2', provided the ratios $A : B : C : D$ be (as they may be) so chosen that three arbitrary points on the perimeter of the one correspond to three arbitrary points on the perimeter of the other. This is the solution of the problem above stated.

In the hydrodynamical application one of the lunes is the strip of the plane of w bounded by the straight lines $\psi = 0$, $\psi = b$. In this case the expression corresponding to the right-hand side of (54), (with w written for z) assumes an indeterminate form; the angular points of the lune being at infinity, whilst its angle is zero. When evaluated by the usual methods this expression becomes $e^{\frac{\pi w}{b}}$. It is in fact obvious from Art. 78 that the assumption

$$Z = X + iY = e^{\frac{\pi w}{b}} \quad \text{.........................(56)}$$

converts the two straight lines in question into the one straight line $Y = 0$, and therefore serves the purpose of (53).

We proceed to give the more important of the applications of the above method which have been made by Kirchhoff.

96. *Example* 10. This is an illustration of the theory of the *vena contracta*. Fluid escapes from a large vessel through an aperture in a plane wall. The forms of the boundaries in the planes of z, ζ, w are shewn in Fig. 7; fixed boundaries being denoted by heavy, free surfaces by fine lines. The figure in the plane of ζ is a lune of angle $\frac{1}{2}\pi$ having its angular points at $\zeta = \pm 1$. The figure in the plane of w has the limiting form just noticed.

Fig. 7.

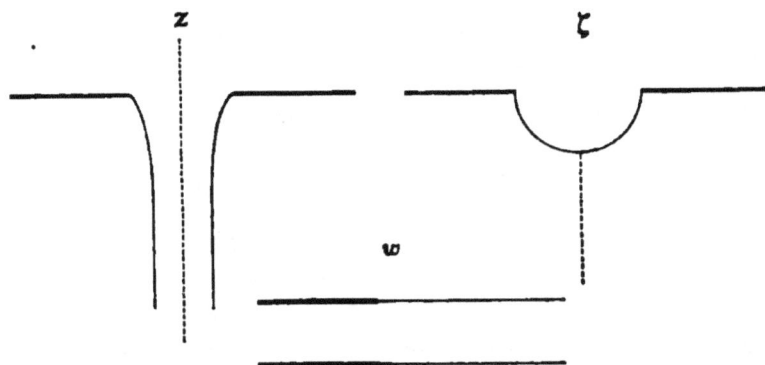

We assume for simplicity that the parameters of the limiting stream-lines are $\psi = 0$, $\psi = \pi$. Applying then the rule developed in the last article, we assume

$$\left(\frac{\zeta - 1}{\zeta + 1}\right)^2 = \frac{Ae^w + B}{Ce^w + D} \dots\dots(57).$$

The arbitrary constants must satisfy the conditions

(*a*) when $\phi = -\infty$, $\zeta = \infty$,

(*b*) when $\phi = +\infty$, $\zeta = -i$;

and if we further take the equipotential surface passing through the edges of the aperture as that for which $\phi = 0$, we must have

(*c*) $w = 0$ when $\zeta = 1$.

Of these conditions (*a*) gives $B = D$, (*b*) gives $A = -C$, and (*c*) gives $A = -B$, so that (57) becomes

$$\left(\frac{\zeta - 1}{\zeta + 1}\right)^2 = \frac{1 - e^w}{1 + e^w} \dots\dots(58),$$

whence we find

$$\frac{dz}{dw} = \zeta = e^{-w} + \sqrt{e^{-2w} - 1} \quad \dots\dots\dots\dots\dots(59).$$

Along the stream-lines $\psi = 0$, $\psi = \pi$, ζ is real so long as ϕ lies between 0 and $-\infty$. To find the form of the free boundaries, let us consider the portion of the stream-line $\psi = 0$ for which $\phi > 0$. The first term of (59) is then real, the second imaginary, so that if we write

$$\zeta = \frac{1}{q} (\cos \theta + i \sin \theta)$$

we have, along the line in question,

$$\cos \theta = e^{-\phi} \dots\dots\dots\dots\dots\dots(60).$$

If s denote the arc of this line, measured from the edge of the aperture, we have

$$\frac{d\phi}{ds} = q = 1,$$

whence $s = \phi$. The equation (60) then gives

$$\frac{dx}{ds} = \cos \theta = e^{-s},$$

and therefore

$$x = 1 + e^{-s} \dots\dots\dots\dots\dots\dots(61),$$

if the origin of z (hitherto arbitrary) be taken at the edge of the aperture. The final width of the stream is given by the difference of the extreme values of ψ (the velocity being unity), i.e. it is equal to π; and since when $s = \infty$, $x = 1$, the abscissa of the centre of the stream is $1 + \frac{1}{2}\pi$. The width of the aperture is therefore $2 + \pi$, and the coefficient of contraction is $\dfrac{\pi}{\pi + 2}$, or ·611.

Again, from (58) or (60) we find

$$\frac{dy}{ds} = \sin \theta = - \sqrt{1 - e^{-2s}},$$

whence

$$y = \sqrt{1 - e^{-2s}} - \tfrac{1}{2} \log \frac{1 + \sqrt{1 - e^{-2s}}}{1 - \sqrt{1 - e^{-2s}}} \quad \dots\dots\dots\dots(62).$$

The equations (61) and (62) combined give the form of the boundary of the issuing jet. They are obtained in the above manner by Lord Rayleigh*, who has also given a drawing of the curve in question.

Since $s = -\log\cos\theta$, the radius of curvature of the boundary is $\dfrac{ds}{d\theta} = \tan\theta$, and therefore vanishes at the edge. Kirchhoff has shewn that this is a general property of free boundaries.

97. *Example* 11. Fluid escapes from a large vessel by a straight canal projecting inwards. This illustrates one of the cases of the tube, spoken of in Art. 31.

An inspection of Fig. 8, giving the forms of the boundaries in the planes of z, w, ζ, and of a new variable $\sqrt{\zeta}$, will shew that

Fig. 8.

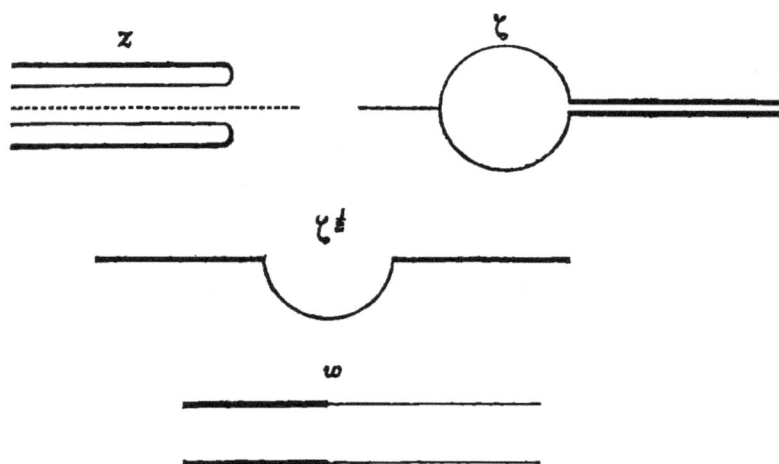

this case may be obtained from the preceding by merely writing $\sqrt{\zeta}$ for ζ, so that we now have

$$\frac{dz}{dw} = \zeta = (e^{-w} + \sqrt{e^{-2w} - 1})^2$$

$$= 2e^{-2w} - 1 + 2e^{-w}\sqrt{e^{-2w} - 1}\ldots\ldots\ldots\ldots(63).$$

* *Phil. Mag.* Dec. 1876.

Along the free boundary $\psi = 0$, $\phi > 0$, we have in the same way as before

$$\frac{dx}{ds} = 2e^{-2s} - 1,$$

or
$$x = 1 - s - e^{-2s},$$

if the origin of x and of s be at the extremity of the fixed wall. Also

$$\frac{dy}{ds} = -2e^{-s}\sqrt{1 - e^{-2s}},$$

whence

$$y = -\tfrac{1}{2}\pi + e^{-s}\sqrt{1 - e^{-2s}} + \arcsin e^{-s},$$

the constant of integration being so chosen as to make $y = 0$ for $s = 0$. When $s = \infty$, $y = -\tfrac{1}{2}\pi$, so that, the final breadth of the stream being as before equal to π, the breadth of the canal is 2π. The coefficient of contraction is therefore $\tfrac{1}{2}$.

This example, the first of its class which was solved, is due to Helmholtz (*l.c.* Art. 92).

If in (58) we write $\zeta^{\frac{\pi}{a}}$ for ζ we obtain the solution of the case where the inclination a of the walls of the canal has any value whatever.

98. *Example* 12. A steady stream impinges directly on a fixed plane lamina. The region of dead water behind the lamina is bounded on each side by a surface of discontinuity at which q is (for the moving fluid) constant (say $= 1$).

The middle stream-line, after meeting the lamina at right angles, branches off into two parts, which follow the lamina to the edges, and thence the surfaces of discontinuity. Let this be the line for which $\psi = 0$, and let us further assume that at the point of divergence we have $\phi = 0$. The forms of the boundaries in the planes of z, ζ, w are shewn in Fig. 9. The region occupied by moving fluid corresponds to the whole of the plane of w; but the two sides of the straight line $\psi = 0$, $\phi > 0$ are internal boundaries. The assumption $w' = \sqrt{w}$ transforms this double line into the axis of abscissæ, which may of course be regarded as a circle of infinite radius.   The rule of Art. 95 then gives

$$\left(\frac{\zeta-1}{\zeta+1}\right)^2 = \frac{A\sqrt{w}+B}{C\sqrt{w}+D} \dots\dots\dots\dots\dots(63).$$

Fig. 9.

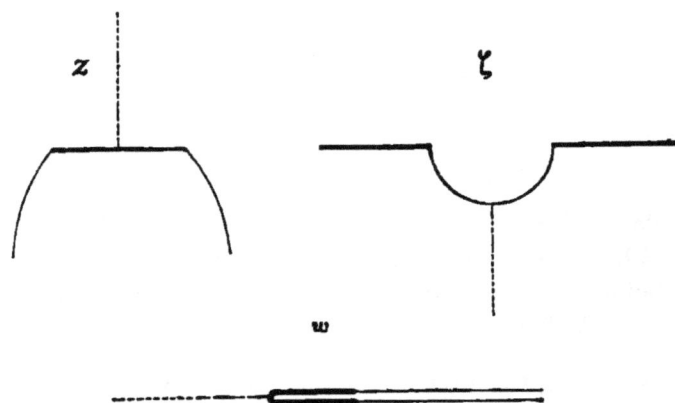

To determine the constants, we have the conditions

$\quad(a)\quad \zeta = -i,\ \text{for}\ \phi = \pm\infty,$

$\quad(b)\quad \zeta = \infty,\ \text{for}\ w = 0;$

to which we add

$\quad(c)\quad \zeta = \pm 1,\ \text{for}\ w = 1.$

The last assumption fixes the breadth of the lamina in terms of the unit of length.   Of these conditions (a) gives $A = -C$, (b) gives $B = D$, and (c) makes $A = -B^*$.   Hence (63) becomes

$$\left(\frac{\zeta-1}{\zeta+1}\right)^2 = \frac{1-\sqrt{w}}{1+\sqrt{w}},$$

or

$$\zeta = \frac{1}{\sqrt{w}} + \sqrt{\frac{1}{w}-1}\dots\dots\dots\dots\dots(64).$$

If $\psi = 0$, and ϕ lie between ± 1, the right-hand side of (64) is

---

* That is, we assume $\sqrt{w} = +1$ when $\zeta = +1$, and therefore $\sqrt{w} = -1$ for $\zeta = -1$. The alternative supposition leads to the same results.

real. To find the breadth l of the lamina in terms of the unit of length, we have

$$l = 2\int_0^1 \frac{dx}{d\phi}\, d\phi = 2\int_0^1 \left(\frac{1}{\sqrt{\phi}} + \sqrt{\frac{1}{\phi}-1}\right) d\phi$$

$$= 4 + \pi \dots\dots\dots\dots\dots \dots\dots\dots(65).$$

To find the excess of pressure on the anterior face of the lamina, we have,

for the moving fluid $\dfrac{p}{\rho} = C - \tfrac{1}{2}q^2$,

and for the dead water $\dfrac{p'}{\rho} = C'$.

Now at the edge of the lamina, we have $p = p'$, $q = 1$, so that $C = C' + \tfrac{1}{2}$, and the required excess is given by

$$\frac{p-p'}{\rho} = \tfrac{1}{2}(1 - q^2).$$

If we multiply this by dx, and integrate between the limits ± 1, since $\dfrac{dx}{d\phi} = \dfrac{1}{q}$, we obtain

$$\int_0^1 \left(\frac{1}{q} - q\right) d\phi,$$

or, since

$$\frac{1}{q} = \frac{1}{\sqrt{\phi}} + \sqrt{\frac{1}{\phi}-1},$$

$$2\int_0^1 \sqrt{\frac{1}{\phi}-1}\, d\phi, \dots\dots\dots\dots\dots(66),$$

which $= \pi$.

To make this result intelligible we must get rid of the arbitrary assumptions made for simplicity of calculation. If q_0 be the general velocity of the stream, ϕ_0 the value of ϕ at the edge of the lamina, we have, instead of (64),

$$q_0\zeta = \sqrt{\frac{\phi_0}{w}} + \sqrt{\frac{\phi_0}{w}-1},$$

so that (65) is replaced by

$$l = (4 + \pi)\frac{\phi_0}{q_0} \dots\dots\dots\dots\dots(67),$$

and (66) by

$$\pi q_0 \phi_0 \dots\dots\dots\dots\dots(68).$$

Multiplying (68) by ρ, and eliminating ϕ_0, we obtain for the total excess of pressure on the anterior face

$$\frac{\pi}{4+\pi} q_0^2 \rho l \dots\dots\dots\dots(69).$$

If the stream be oblique to the lamina, making an angle α with its plane, the condition (a) is replaced by

$$\zeta = e^{-i\alpha}, \text{ for } \phi = \pm \infty,$$

which gives

$$A : C = \cos \alpha - 1 : \cos \alpha + 1;$$

(b) is unaltered, whilst (c) is no longer applicable, there being no longer symmetry as regards the line $\psi = 0$. The former conditions however reduce (63) to the form

$$\left(\frac{\zeta-1}{\zeta+1}\right)^2 = \frac{K-(1-\cos\alpha)\sqrt{w}}{K+(1+\cos\alpha)\sqrt{w}},$$

which shews that the value of the remaining constant K only affects the scale of w. If we assign to it any real value, we make the cusp $\zeta = 1$ of the lune in the plane of ζ correspond to some definite point of the boundary in the plane of w'. The simplest assumption is $K = 1$, which gives, after some reduction,

$$\zeta = \cos \alpha + \frac{1}{\sqrt{w}} + \sqrt{\left(\cos \alpha + \frac{1}{\sqrt{w}}\right)^2 - 1}.$$

For the discussion of this result, and the calculation of the resultant pressure on the lamina we must refer to the paper by Lord Rayleigh, already cited (*l. c.* Art. 96).

CHAPTER V.

99. THE chief subject treated of in this chapter is the motion of a solid through an infinite mass of liquid under the action of any given forces. The same analysis applies with little or no alteration to the case of a liquid occupying a cavity in a moving solid. We shall consider, though less fully, cases where we have more than óne moving solid, or where the fluid does not extend in all directions to infinity, being bounded externally by fixed rigid walls.

We shall assume in the first instance that the motion of the fluid is entirely due to that of the solid, and is therefore characterized by the existence of a single-valued velocity-potential ϕ which besides satisfying the equation of continuity

$$\nabla^2\phi = 0 \dots\dots\dots\dots\dots\dots\dots\dots\dots\dots(1)$$

fulfils the following conditions: (a) the value of $\dfrac{d\phi}{dn}$, dn denoting as usual an element of the normal at any point of the surface of the solid drawn towards the fluid, must be equal to the velocity of the surface at that point normal to itself, and (b) the differential coefficients $\dfrac{d\phi}{dx}$, $\dfrac{d\phi}{dy}$, $\dfrac{d\phi}{dz}$ must vanish at an infinite distance, in every direction, from the solid. The latter condition is rendered necessary by the consideration that a finite velocity at infinity would imply an infinite kinetic energy, which could not be generated by finite forces acting for a finite time on the solid. It is also the condition to which we are led by supposing the fluid enclosed within a fixed vessel infinitely large and infinitely distant all round from the moving body. For on this supposition the space occupied by the fluid may be conceived as made up of tubes of flow which begin and end on the surface of the solid, so that the total flux across any area, finite or infinite, drawn in the fluid must be finite, and therefore the velocity at infinity zero.

In the case of a fluid occupying a cavity in a moving solid, the condition (b) does not apply; the surface of the cavity is then the complete boundary of the fluid, and the condition (a) is therefore sufficient.

We have seen in Arts. 49, 52 that under either of the above sets of conditions the motion of the fluid is determinate. Our problem then divides itself into two distinct parts; the first or kinematical part consisting in the determination of the motion of the fluid at any instant in terms of that of the solid, and the second, or dynamical part, in the calculation of the effect of the fluid pressures on the surface of the latter.

Kinematical Investigations.

100. Let us take a system of rectangular axes Ox, Oy, Oz fixed in the body, and let the motion of the latter at any time t be defined by the instantaneous angular velocities p, q, r about, and the translational velocities u, v, w of the origin O parallel to, the instantaneous positions of these axes. We may then write, as Kirchhoff does,

$$\phi = u\phi_1 + v\phi_2 + w\phi_3 + p\chi_1 + q\chi_2 + r\chi_3 \cdots\cdots\cdots(2),$$

where, as will immediately appear, ϕ_1, &c., χ_1, &c., are certain functions of x, y, z depending only on the shape and size of the solid. In fact, if l, m, n denote the direction-cosines of the normal (drawn on the side of the fluid) at any point (x, y, z) of the surface of the solid, the condition (a) of Art. 99 may be written

$$\frac{d\phi}{dn} = (u + qz - ry)\,l + (v + rx - pz)\,m + (w + py - qx)\,n\,;$$

and equating this to the value of $\dfrac{d\phi}{dn}$ obtained from (2) we find

$$\left.\begin{array}{lll} \dfrac{d\phi_1}{dn} = l, & \dfrac{d\phi_2}{dn} = m, & \dfrac{d\phi_3}{dn} = n, \\[2mm] \dfrac{d\chi_1}{dn} = ny - mz, & \dfrac{d\chi_2}{dn} = lz - nx, & \dfrac{d\chi_3}{dn} = mx - ly. \end{array}\right\} \cdots\cdots(3).$$

The functions ϕ_1, &c. must of course separately satisfy (1), and have their derivatives zero at infinity; the surface-conditions (3) then render them completely determinate.

101.   When the motion is in two dimensions (xy) we have only three functions to determine, viz. ϕ_1, ϕ_2, χ_3. In the last chapter (Arts. 88, 89) general methods for discovering cases in which *one* of these functions is known were given. In any case of a liquid filling a cavity in a moving solid it is plain that the conditions (3) are satisfied by ϕ_1, ϕ_2, $\phi_3 = x$, y, z, respectively, in other words that if the solid have a motion of translation only, the enclosed fluid moves as if it formed a rigid mass. We may therefore regard the kinematical part of our problem as solved for the cases where the cavity is in the form of an elliptic cylinder, or a triangular prism on an equilateral base, for which χ_3 has been found*.

In the more difficult problem of a cylindrical body moving through an infinite mass of liquid, the complete solution has been obtained for the case where the section of the cylinder is elliptic, and for this case only.

102.   The number of cases in three dimensions for which the functions ϕ_1, &c., χ_1, &c. have been completely determined is very small. We give here the chief of them.

Example 1.   An ellipsoidal cavity whose semiaxes are a, b, c. If the principal axes of the ellipsoid be taken as axes of co-ordinates, and h be the perpendicular from the centre on the tangent plane at (x, y, z), we have then

$$l, \; m, \; n = \frac{hx}{a^2}, \; \frac{hy}{b^2}, \; \frac{hz}{c^2},$$

respectively, so that the last three of conditions (3) give

$$\frac{d\chi_1}{dn} = h\left(\frac{1}{c^2} - \frac{1}{b^2}\right) yz, \text{ &c., &c.}$$

But also

$$\frac{d\chi_1}{dn} = \frac{d\chi_1}{dx} l + \frac{d\chi_1}{dy} m + \frac{d\chi_1}{dz} n.$$

* The student will find in Thomson and Tait, *Natural Philosophy*, Art. 708, other forms of cylindrical cavity for which solutions can be obtained.

These two values of $\frac{d\chi_1}{dn}$ agree provided we put $\chi_1 = Ayz$, and make

$$A\left(\frac{1}{b^2} + \frac{1}{c^2}\right) = \frac{1}{c^2} - \frac{1}{b^2}.$$

We have then, finally,

$$\phi = ux + vy + wz$$
$$+ p\,\frac{b^2 - c^2}{b^2 + c^2}yz + q\,\frac{c^2 - a^2}{c^2 + a^2}zx + r\,\frac{a^2 - b^2}{a^2 + b^2}xy \dots\dots(4).$$

Example 2.   A cavity in the form of a rectangular parallel-epiped. If the axes of co-ordinates be taken parallel to the edges of the cavity, it is plain that the conditions (3) are satisfied by making χ_1 a function of y, z only, &c., so that the problem becomes one of two dimensions. For the complete solution, effected by means of Fourier's series, we refer the student to Stokes[*], or to Thomson and Tait[†].

Example 3.   An ellipsoid moves in an infinite mass of liquid which is at rest at infinity.

This problem, the only one of its class which has been completely worked out, was solved by Green[‡] in 1833, for the particular case where the motion of the ellipsoid is one of pure translation. The complete solution was published by Clebsch[§], in 1856; it is here reproduced much in the form given to it by Kirchhoff[||].

The principal axes of the ellipsoid being taken as axes of co-ordinates, let the component attractions which would be exerted at the point (x, y, z) by an ellipsoid of unit density, coincident in shape, size, and position with the given one, be denoted by X, Y, Z. It is known that X is the potential of the ellipsoid when magnetized uniformly with unit intensity parallel to x negative, and therefore that it is the potential of a distribution of matter, of surface-density $- l$, over the surface[¶]. Hence $\frac{dX}{dn}$ is discon-

[*] *Camb. Phil. Trans.*, Vol. VIII., pp. 131, 409.

[†] *Natural Philosophy*, Art. 707 (B).

[‡] 'Researches on the Vibration of Pendulums in Fluid Media,' *Trans. R.S. Edin.* 1833. Reprinted in Mr Ferrers' edition of Green's works, pp. 315 *et seq.*

[§] *Crelle*, tt. 52, 53 (1856—7).

[||] *Vorlesungen über Math. Physik. Mechanik.*, c. 18.

[¶] See, as to these points, Maxwell, *Electricity and Magnetism*, Art. 437.

tinuous at the surface; viz. distinguishing by $\left[\dfrac{dX}{dn}\right]$ and $\left\{\dfrac{dX}{dn}\right\}$ its values just outside, and just inside, respectively, we have

$$\left[\frac{dX}{dn}\right] - \left\{\frac{dX}{dn}\right\} = 4\pi l \dots\dots\dots\dots(5).$$

But at an internal point we have (Thomson and Tait, Art. 522),

$$X = -Fx \dots\dots\dots\dots\dots(6),$$

where

$$F = 2\pi abc \int_0^\infty \frac{d\lambda}{(a^2 + \lambda)^{\frac{3}{2}} (b^2 + \lambda)^{\frac{1}{2}}(c^2 + \lambda)^{\frac{1}{2}}} \dots\dots\dots(7),$$

a, b, c being the semiaxes of the ellipsoid.   Hence

$$\left\{\frac{dX}{dn}\right\} = -Fl,$$

so that (5) gives

$$\left[\frac{dX}{dn}\right] = (4\pi - F)l \dots\dots\dots\dots(8).$$

Since X of course satisfies (1), and has its derivatives zero at infinity, it is plain that all the conditions of the question are satisfied by

$$\phi_1 = \frac{1}{4\pi - F} X.$$

The value of X at an external point (x, y, z) is (Thomson and Tait, *l. c.*),

$$X = -2\pi abcx \int_\lambda^\infty \frac{d\lambda}{(a^2 + \lambda)^{\frac{3}{2}}(b^2 + \lambda)^{\frac{1}{2}}(c^2 + \lambda)^{\frac{1}{2}}} \dots\dots(9),$$

where the lower limit is the positive root of

$$\frac{x^2}{a^2 + \lambda} + \frac{y^2}{b^2 + \lambda} + \frac{z^2}{c^2 + \lambda} = 1.$$

We have, in exactly the same way,

$$\phi_2 = \frac{1}{4\pi - G} Y, \qquad \phi_3 = \frac{1}{4\pi - H} Z,$$

where the values of G, H, Y, Z may be written down from (7) and (9) by symmetry.

Next let us consider the function

$$U = yZ - zY \dots\dots\dots\dots\dots\dots(10).$$

With the same notation as before we find

$$\left[\frac{dU}{dn}\right] = mZ - nY + y\frac{dZ}{dn} - z\frac{dY}{dn}$$

$$= ny(4\pi - H + G) - mz(4\pi - G + H),$$

by formulæ of the same type as (6) and (8). Since

$$m : n = \frac{y}{b^2} : \frac{z}{c^2},$$

the ratio of the expression last written to $ny - mz$ is

$$\frac{b^2(4\pi - H + G) - c^2(4\pi - G + H)}{b^2 - c^2},$$

a constant. Hence all the conditions of the problem are satisfied by making

$$\chi_1 = \frac{b^2 - c^2}{4\pi(b^2 - c^2) + (G - H)(b^2 + c^2)} U,$$

where

$$U = -2\pi abc(b^2 - c^2)yz \int_\lambda^\infty \frac{d\lambda}{(a^2+\lambda)^{\frac{3}{2}}(b^2+\lambda)^{\frac{3}{2}}(c^2+\lambda)^{\frac{3}{2}}} \dots\dots(11),$$

the lower limit being the same as in (9).

The values of χ_2, χ_3 may be written down from symmetry.

The student may, as an exercise, prove the equivalence of the above formulæ, in the case where one of the axes of the ellipsoid is infinite, to those of Arts. 88 (d) and 89 (c).

Example 4. In the particular case of a sphere we have $a = b = c$. We then find

$$F = \tfrac{4}{3}\pi, \qquad X = -\tfrac{4}{3}\frac{\pi a^3 x}{r^3},$$

(where $r^2 = x^2 + y^2 + z^2$), and therefore

$$\phi_1 = -\tfrac{1}{2}\frac{a^3 x}{r^3},$$

with similar formulæ for ϕ_2, ϕ_3. The values of χ_1, χ_2, χ_3 are zero, as is obvious *a priori*.

The solution in this case may however be obtained *ab initio* by a simpler analysis, as follows[*].

Let OX be the direction of motion of the centre O of the sphere at any instant, and V its velocity. Let P be any point of the fluid, and let $OP = r$, and the angle $POX = \theta$. It is evident that ϕ will be a function of r, θ only; and we know from the theory of Spherical Harmonics that any such function which satisfies (1) and has its derivatives zero at infinity can be expanded in a series of the form

$$\phi = \text{const.} + \frac{Q_0}{r} + \frac{Q_1}{r^2} + \frac{Q_2}{r^3} + \&c.,$$

where Q_n is the 'zonal harmonic' of order n, multiplied by an arbitrary constant. The condition which ϕ has to satisfy at the surface of the sphere is

$$\frac{d\phi}{dr} = V \cos \theta,$$

so that, if a be the radius,

$$V \cos \theta = -\frac{Q_0}{a^2} - \frac{2Q_1}{a^3} - \frac{3Q_2}{a^4} - \&c.$$

Hence

$$-\frac{2Q_1}{a^3} = V \cos \theta,$$

and

$$Q_0 = Q_2 = Q_3 = \&c. = 0.$$

We have then finally

$$\phi = \text{const.} - \tfrac{1}{2} \frac{Va^3}{r^2} \cos \theta \dots \dots \dots \dots \dots (12).$$

It is easy to verify the fact that this value of ϕ really satisfies all the conditions of the problem.

If we impress on the whole system—the moving sphere and the fluid—a velocity $-V$, in the direction OX, we have the case of a uniform stream of velocity V flowing past a fixed spherical obstacle. The velocity-potential is then got by adding the term $-Vr \cos \theta$ to (12), so that we now have

$$\phi = \text{const.} - V\left(r + \tfrac{1}{2}\frac{a^3}{r^2}\right) \cos \theta.$$

---

[*] This solution, generally attributed by continental writers to Dirichlet (*Monatsberichte der Berl. Akad.* 1852), was given by Stokes, *Camb. Trans.* Vol. VIII. (1843).

103.　A method similar to that of Art. 88 has been employed by Rankine* to discover forms of solids of revolution (about x say) for which ϕ_1 is known.　When such a solid moves parallel to its axis the motion generated in the fluid takes place in a series of planes through that axis and is the same in each such plane. In all cases of motion of this kind there exists a stream-function analogous to that of Chapter IV.　If we take in any plane through Ox two points A and P, A fixed and P variable, and consider the annular surface generated by the revolution about x of any line AP, it is plain that the quantity of fluid which in unit time crosses this surface is a function of the position of P, i.e. it is a function of x and ϖ, where ϖ denotes the distance of P from Ox.　Let this function be denoted by $2\pi\psi$.　The curves $\psi = $ const. are evidently stream-lines, so that ψ may be called the 'stream-function.'　If P' be a point infinitely close to P in the above-mentioned plane, we have from the definition of ψ

$$\text{fluid velocity normal to } PP' = \frac{2\pi d\psi}{2\pi\varpi \,.\, PP'},$$

and thence, taking PP' parallel, first to ϖ, then to x,

$$u = \frac{1}{\varpi}\frac{d\psi}{d\varpi}, \qquad v = -\frac{1}{\varpi}\frac{d\psi}{dx} \quad\ldots\ldots\ldots\ldots\ldots(13),$$

where u, v are the components of fluid velocity parallel to x and ϖ respectively.

For the case of the sphere, treated in Art. 102, we readily find, by comparison of (12) and (13)

$$\psi = \tfrac{1}{2}Va^3\,\frac{\varpi^2}{r^3} = \tfrac{1}{2}Va^3\,\frac{\sin^2\theta}{r} \quad\ldots\ldots\ldots\ldots\ldots(14).$$

So far we have not assumed the motion to be irrotational. The condition that it should be so, is

$$\frac{du}{d\varpi} - \frac{dv}{dx} = 0,$$

which reduces to

$$\frac{d^2\psi}{dx^2} + \frac{d^2\psi}{d\varpi^2} - \frac{1}{\varpi}\frac{d\psi}{d\varpi} = 0 \quad\ldots\ldots\ldots\ldots(15).$$

* *Phil. Trans.* 1871.

The differential equation satisfied by ϕ now assumes the form

$$\frac{d^2\phi}{dx^2} + \frac{d^2\phi}{d\varpi^2} + \frac{1}{\varpi}\frac{d\phi}{d\varpi} = 0 \quad \ldots\ldots\ldots\ldots\ldots\ldots(16).$$

This may be derived by transformation from (1), by writing

$$y = \varpi \cos\theta, \quad z = \varpi \sin\theta,$$

and remembering that ϕ is now independent of θ, or by repeating the investigation of Art. 12, taking, instead of the elementary volume $dx\,dy\,dz$ there considered, the annular space generated by the revolution about x of the rectangle $dx\,d\varpi$. It appears that ϕ and ψ are not, as they were in Chapter IV., interchangeable.

104.   Rankine's procedure is then as follows. Supposing the solid to move parallel to its axis with velocity V, we have at all points of a section of its surface made by a plane through Ox,

$$\frac{d\psi}{\varpi ds} = \text{velocity in direction of normal}$$

$$= V\frac{d\varpi}{ds},$$

ds denoting an element of the said section.

Integrating along the section, we find

$$\psi = \tfrac{1}{2}V\varpi^2 + \text{const.} \quad \ldots\ldots\ldots\ldots\ldots\ldots\ldots\ldots(17).$$

If in this equation we substitute any value of ψ satisfying (15), we obtain the equation of the meridional section of a series of solids of revolution, any one of which would when moving parallel to its axis produce the system of stream-lines corresponding to the assumed value of ψ.

In this way may be verified the value (14) of ψ for the case of a sphere.

Dynamical Investigations.

105.   The second part of the problem proposed in Art. 99 is the determination of the effect of the fluid pressure on the

surface of the solid. The most obvious way of doing this is, first to calculate p from the formula

$$\frac{p}{\rho} = \text{const.} - \frac{d\phi}{dt} - \tfrac{1}{2} \, (\text{vel.})^2 \, \dots\dots\dots\dots(18),$$

and then to find the resultant force and couple due to the pressure p acting on the various elements dS of the surface by the ordinary rules of Statics. We will work out the result for the simple case of the sphere, starting from the value of ϕ given by (12). Since the origin to which ϕ·is there referred is in motion parallel to OX with velocity V, whereas in (18) the origin is supposed fixed, we must write, instead of $\dfrac{d\phi}{dt}$,

$$\frac{d\phi}{dt} - V \frac{d\phi}{dx},$$

where $x = r \cos \theta$. Now

$$\frac{d\phi}{dx} = \frac{d\phi}{dr} \cos \theta - \frac{d\phi}{rd\theta} \sin \theta = \frac{Va^3}{r^3} (\cos^2 \theta - \tfrac{1}{2} \sin^2 \theta),$$

and

$$(\text{vel.})^2 = \left(\frac{d\phi}{dr}\right)^2 + \left(\frac{d\phi}{rd\theta}\right)^2 = \frac{V^2 a^6}{r^6} (\cos^2 \theta + \tfrac{1}{4} \sin^2 \theta).$$

The whole effect of the fluid pressure evidently reduces to a force in the direction OX. The value of p at the surface of the sphere is

$$p = \text{const.} + \tfrac{1}{2}\rho a \frac{dV}{dt} \cos \theta + \&\text{c.},$$

the remaining terms being the same for surface-elements in the positions θ and $\pi - \theta$, and therefore not affecting the final result. Hence if V be constant, the pressures on the various elements of the anterior half of the sphere are balanced by equal pressures on the corresponding elements of the posterior half; but when the motion of the sphere is being accelerated there is an excess of pressure on the anterior, and a defect of pressure on the posterior half. The reverse holds when the motion is being retarded. The total effect in the direction of V, is

$$- \int_0^\pi 2\pi a \sin \theta \,.\, ad\theta \,.\, p \cos \theta,$$

which is readily found to be equal to $-\frac{2}{3}\pi\rho a^3 \dfrac{dV}{dt}$, or $-\frac{1}{2}M' \dfrac{dV}{dt}$, if M' denote the mass of fluid displaced by the sphere.

If we suppose that the sphere started from rest under the action of a force X constant in direction, so that the centre moves in a straight line, the equation is

$$M\frac{dV}{dt} = X - \tfrac{1}{2}M' \frac{dV}{dt},$$

or
$$(M + \tfrac{1}{2}M')\frac{dV}{dt} = X.\dotfill(19).$$

The sphere therefore behaves exactly as if its inertia were increased by half that of the fluid displaced, and the surrounding fluid were annihilated.

We have assumed throughout the above calculation that the motion of the sphere is rectilinear. It is not difficult to extend the result to the case where the motion is of any kind whatever. This is effected however more simply by the method of the next article.

The same method can be applied with even greater ease to the case of a long circular cylinder, for which the value of ϕ was obtained in Art. 86. It appears that the effect of the fluid pressure is in that case to increase the inertia of the cylinder by that of the fluid displaced exactly.

The practical value of these results, and of similar more general ones to be obtained below, is discussed in note (E).

106.   The above direct method of calculating the forces exerted by the fluid on the moving body would, however, in most cases prove exceedingly tedious. This difficulty may be avoided by a method, first used by Thomson and Tait[*], which consists in treating the solid and the fluid as forming together one dynamical system, into the equations of motion of which the mutual reactions of the solid and the fluid of course do not enter. As a

---

[*] *Natural Philosophy*, Art. 331.

very simple example of this method we may take the case of the rectilinear motion of a sphere, which has been already investigated otherwise. By the formula of Art. 65, the kinetic energy of the fluid

$$= - \tfrac{1}{2}\rho \iint \phi \, \frac{d\phi}{dn} \, dS,$$

which in our case

$$= - \tfrac{1}{2}\rho \int_0^\pi \phi \, \frac{d\phi}{dr} \cdot 2\pi a \sin \theta \cdot a \, d\theta$$

$$= \tfrac{1}{2}\pi\rho a^3 V^2 \int_0^\pi \cos^2 \theta \cdot \sin \theta \, d\theta,$$

by (12), or finally $\tfrac{1}{2}M' V^2$. Hence the total energy of the system is

$$\tfrac{1}{2} \left(M + \tfrac{1}{2}M' \right) V^2.$$

The rate at which this is increasing, $i.e.$

$$\left(M + \tfrac{1}{2}M' \right) V \frac{dV}{dt},$$

must be equal to XV, the rate at which the impressed force X does work. Discarding the common factor V we are led again to the equation (19).

107. In the general case the motion of the fluid at any instant depends, as we saw in Art. 100, only on the values of the quantities u, v, w, p, q, r used to express the motion of the solid; so that the whole dynamical system is virtually one of six degrees of freedom, although it differs in some respects from the kind of system ordinarily contemplated in Dynamics. Thomson* and Kirchhoff† have independently shewn how a system of the peculiar kind here considered may be brought under the application of the ordinary methods of that science. We shall, in what follows, adopt Thomson's procedure, with some modifications.

Whatever be the motion of the fluid and solid at any instant, we may suppose it produced instantaneously from rest by the

---

* $Phil. Mag.$ November, 1871.

† $Crelle$, t. 71. See also $Vorlesungen~über~Math.~Physik.~Mechanik.$ c. 19.

action of a properly chosen set of impulsive forces applied to the solid. This set, when reduced after the manner of Poinsot to a force and a couple whose axis is parallel to the line of action of the force, constitute what Thomson calls the 'impulse' of the motion at the instant under consideration. We proceed to shew that when no external impressed forces act the impulse is constant in every respect throughout the motion.

108.   The moment of momentum of a spherical portion of the fluid about any line through its centre is zero; for this portion may be conceived as made up of circular rings of infinitely small section having this line as a common axis, and the circulation in each such ring is zero.

In the same way the moment of momentum of a portion of the fluid bounded by two spherical surfaces about the line joining the centres is zero.

The moment of the impulse at any instant about any line is equal to the corresponding moment of momentum at that instant of the whole matter contained within a spherical surface having its centre in that line and enclosing the moving solid; for if we suppose the motion generated instantaneously from rest, the only forces which, besides those constituting the impulse, act on the mass in question are the impulsive pressures on the spherical boundary. Since these act in lines through the centre, they do not affect the moment of momentum.

It is, as was pointed out in Art. 99, immaterial whether we simply suppose the fluid to extend to infinity and to be at rest there, or whether we suppose it contained in an infinitely large fixed rigid vessel which is infinitely distant in all directions from the moving solid. The motion of the fluid within a finite distance of the solid, and therefore the forces exerted by it on the latter, are the same in the two cases. If we now suppose the infinite containing vessel to be spherical in shape, and to have its centre at any point P within a finite distance of the solid, the moment of momentum of the included mass about any line through P is, as we have just seen, equal to the moment of the impulse about the same line. The same reasoning shews that if there be no external impressed forces this moment of momentum is constant throughout the motion. Hence the moment of the impulse about any line through P is constant. Since in this argument P may be any point within a finite distance of the solid, it follows that the moment of the impulse about any line whatever is constant. This cannot be the case unless the impulse is itself constant in every respect.

We see in the same way that if any external impressed forces act on the solid, the moment of the impulse about any line is increasing at any instant at a rate equal to the moment of these forces about the same line.

The above are somewhat modified proofs of theorems first given by Thomson[*]. It should be noticed that the reasoning still holds when the single solid is replaced by a group of solids, which may moreover (if of invariable volume) be flexible instead of rigid, and even when these solids are replaced by portions of fluid moving rotationally.

109. The 'impulse' then varies in consequence of the action of the external impressed forces in exactly the same way as the momentum of any ordinary dynamical system does. To express this result analytically let $\xi,\ \eta,\ \zeta;\ \lambda,\ \mu,\ \nu$ denote the components of the force- and couple-constituents of the impulse; and let $X,\ Y,\ Z;\ L,\ M,\ N$ designate in the same manner the system of external impressed forces. The whole variation of $\xi,\ \eta,\ \zeta,$ &c., due partly to the motion of the axes to which these quantities are referred, and partly to the action of the forces $X,\ Y,\ Z,$ &c., is then given by the formulæ[†]:

$$
\left.
\begin{aligned}
\frac{d\xi}{dt} &= r\eta - q\zeta + X, & \frac{d\lambda}{dt} &= w\eta - v\zeta + r\mu - q\nu + L, \\
\frac{d\eta}{dt} &= p\zeta - r\xi + Y, & \frac{d\mu}{dt} &= u\zeta - w\xi + p\nu - r\lambda + M, \\
\frac{d\zeta}{dt} &= q\xi - p\eta + Z, & \frac{d\nu}{dt} &= v\xi - u\eta + q\lambda - p\mu + N.
\end{aligned}
\right\} \ \dots\dots(20).
$$

[*] On Vortex Motion.
[†] See Hayward, *Camb. Trans.* Vol. x.

When no external forces act, these equations have the first integrals

$$\xi^2 + \eta^2 + \zeta^2 \qquad\qquad\qquad = \text{const.},$$

$$\lambda\xi + \mu\eta + \nu\zeta \qquad\qquad\qquad = \text{const.},$$

$$u\xi + v\eta + w\zeta + p\lambda + q\mu + r\nu = 2T = \text{const.},$$

of which the first and second together express the fact that the magnitudes of the force- and couple-constituents of the impulse are constant, and the third the fact that the whole energy of the motion is constant.

110. It remains to express ξ, η, ζ, &c. in terms of u, v, w, &c. In the first place let $\mathfrak{T}$ denote the kinetic energy of the fluid alone, so that

$$2\mathfrak{T} = -\rho \iint \phi \frac{d\phi}{dn} dS,$$

where the integration extends over the whole surface of the moving solid. Substituting in this formula the value (2) of ϕ, we get for $2\mathfrak{T}$ an expression of the form

$$\begin{aligned}
2\mathfrak{T} = {} & Au^2 + Bv^2 + Cw^2 + 2A'vw + 2B'wu + 2C'uv \\
& + Pp^2 + Qq^2 + Rr^2 + 2P'qr + 2Q'rp + 2R'pq \\
& + 2p(Lu + Mv + Nw) \\
& + 2q(L'u + M'v + N'w) \\
& + 2r(L''u + M''v + N''w) \dots\dots\dots\dots\dots\dots\dots(21),
\end{aligned}$$

where A, B, C, &c. are certain constant coefficients whose values depend only on the form of the solid, and on the position of the axes of co-ordinates relative to the solid; viz. we have

$$\begin{aligned}
A = {} & -\rho \iint \phi_1 \frac{d\phi_1}{dn} dS = -\rho \iint \phi_1 l \, dS, \\[4pt]
A' = {} & -\tfrac{1}{2}\rho \iint \left(\phi_2 \frac{d\phi_3}{dn} + \phi_3 \frac{d\phi_2}{dn} \right) dS \\[4pt]
& = -\rho \iint \phi_2 \frac{d\phi_3}{dn} dS = -\rho \iint \phi_3 \frac{d\phi_2}{dn} dS \\[4pt]
& = -\rho \iint \phi_2 n \, dS = -\rho \iint \phi_3 m \, dS, \\[4pt]
P = {} & -\rho \iint \chi_1 \frac{d\chi_1}{dn} dS = -\rho \iint \chi_1 (ny - mz) \, dS, \\
& \quad\text{\&c.,} \qquad \text{\&c.,} \qquad \text{\&c.}
\end{aligned} \right\} \dots\dots(22),$$

the transformations being effected by the use of (3) and of a particular case of Green's theorem. These expressions for the coefficients are due to Kirchhoff.

The kinetic energy, $\mathfrak{T}'$ say, of the solid alone is also given by a quadratic function of u, v, w, &c., in which however A, B, C are each equal to the mass of the solid, whilst A', B', C', L, M, N, &c. all vanish. The total energy $\mathfrak{T} + \mathfrak{T}'(= T$, say,) of the system is therefore given by a formula of the same form as (21). Except when otherwise indicated we shall suppose A, B, C, &c. to stand for the coefficients in the expression for twice this total energy.

111. The only form of solid for which the coefficients in the expression (21) for $2\mathfrak{T}$ have been actually determined is the ellipsoid. We readily find

$$A = \frac{F}{4\pi - F} \cdot \tfrac{4}{3} \pi \rho abc,$$

$$P = \tfrac{1}{5} \cdot \frac{(b^2 - c^2)^2 (H - G)}{4\pi (b^2 - c^2) + (G - H)(b^2 + c^2)} \cdot \tfrac{4}{3} \pi \rho abc,$$

where the notation is the same as in Art. 102, Ex. 3. The values of B, C, Q, R may be written down from symmetry; those of the remaining coefficients are all zero. See Art. 116 (d). Since

$$A - B = \frac{4\pi (F - G)}{(4\pi - F)(4\pi - G)} \cdot \tfrac{4}{3} \pi \rho abc,$$

it appears that if $a > b > c$, then $A < B < C$, as might have been anticipated.

112. When in any dynamical system the expression for the kinetic energy in terms of the velocities is known, the values of the component momenta can be derived by a perfectly general process. For this we must refer to books on general Dynamics*. Applied to our case it gives

$$\xi,\ \eta,\ \zeta,\ \lambda,\ \mu,\ \nu = \frac{dT}{du},\ \frac{dT}{dv},\ \frac{dT}{dw},\ \frac{dT}{dp},\ \frac{dT}{dq},\ \frac{dT}{dr} \dots (23),$$

respectively. These formulæ are readily deduced from those which relate to a perfectly free rigid body by supposing the

---

* See Thomson and Tait, *Nat. Phil.* Art. 313, or Maxwell, *Electricity and Magnetism*, Part IV. c. 5. An outline of the process adapted to our case is given in note (C).

motion at any instant generated impulsively from rest, and calculating the effect of the impulsive fluid pressures on the surface of the solid. For instance, the resultant impulsive force parallel to x due to this cause is

$$\iint \rho \phi \, l \, dS, \quad \text{or} \quad \rho \iint \phi \, \frac{d\phi_1}{dn} \, dS,$$

i.e. by (2) and (22),

$$- (Au + C'v + B'w + Lp + L'q + L''r),$$

(if A, B, C, &c. be supposed for a moment to refer to the fluid only), or $-\dfrac{d\mathfrak{T}}{du}$. Hence

$\dfrac{d\mathfrak{T}'}{du} =$ momentum of solid parallel to x, (by ordinary Dynamics)

     $=$ total impulse in same direction

     $= \xi - \dfrac{d\mathfrak{T}}{du}$;

so that

$$\xi = \frac{d \, (\mathfrak{T} + \mathfrak{T}')}{du} = \frac{dT}{du} ;$$

and in the same way the rest of the formulæ (23) may be verified.

113. The equations of motion (20) may now be written in the form

$$\begin{aligned}
&\frac{d}{dt}\frac{dT}{du} = r\frac{dT}{dv} - q\frac{dT}{dw} + X, \\
&\qquad \text{\&c.,} \qquad \text{\&c.,} \\
&\frac{d}{dt}\frac{dT}{dp} = w\frac{dT}{dv} - v\frac{dT}{dw} + r\frac{dT}{dq} - q\frac{dT}{dr} + L, \\
&\qquad \text{\&c.,} \qquad \text{\&c.}
\end{aligned} \right\} \dots\dots(24).$$

We can at once derive some interesting conclusions from these equations, in the case where no external forces act. In the first place Kirchhoff has pointed out that (24) are then satisfied by p, q, $r = 0$, and u, v, w constant, provided we have

$$\frac{dT}{du} : u = \frac{dT}{dv} : v = \frac{dT}{dw} : w ;$$

i.e. provided the velocity of which *u, v, w* are the components be in the direction of one of the principal axes of the ellipsoid,

$$Ax^2 + By^2 + Cz^2 + 2A'yz + 2B'zx + 2C'xy = \text{const.}$$

There exist then for every body three mutually perpendicular directions of permanent translation; that is to say, if the body be set in motion parallel to one of these directions, without rotation, and then left to itself, it will continue to move in this manner. It will be seen that these directions are determined by the ratio of the mean density of the solid to the density of the surrounding fluid and by the *form* of the body's surface. The impulse necessary to produce motion in one of these directions does not in general reduce to a single force; thus if the axes of co-ordinates be chosen, for convenience, parallel to these directions, so that A', B', $C' = 0$, we have corresponding to the motion *u* alone

$$\xi = Au, \quad \eta = 0, \quad \zeta = 0 ;$$
$$\lambda = Lu, \quad \mu = 0, \quad \nu = 0 ;$$

so that the impulse consists of a wrench* of pitch $\dfrac{L}{A}$.

114. The above, although the simplest, are not the only steady motions of which the body is capable (under the action of no external forces). The instantaneous motion of the body at any instant consists, by a well-known theorem of Kinematics, of a twist about a certain screw†; and the condition that this motion should be permanent is that it should not affect the configuration of the impulse (which is fixed in space) relatively to the body. This requires that the axes of the screw and of the corresponding impulsive wrench should coincide. Since the general equations of a straight line involve four independent constants, this gives four relations to be satisfied by the five ratios $u : v : w : p : q : r$.

---

* A 'wrench' is a system of forces supposed reduced after the manner of Poinsot to a force and a couple having its axis in the direction of the force. Its 'pitch' is the line which is the result of dividing the couple by the force. See Ball, *Theory of Screws.*

† A 'twist' is the most general motion of a rigid body, equivalent to a translation parallel to some axis combined with a rotation about that axis. Its 'pitch' is the linear magnitude which is the ratio of the translation to the rotation. Ball, *Theory of Screws.*

There exists then for every body, under the circumstances here considered, a simply-infinite system of possible steady motions.

Of these the next in importance to the three motions of permanent translation are those in which the impulse reduces to a couple only. The equations (20) or (24) are satisfied by ξ, η, $\zeta = 0$, and λ, μ, ν constant, provided

$$\frac{\lambda}{p} = \frac{\mu}{q} = \frac{\nu}{r}, \quad = k, \text{ say} \dots\dots\dots\dots(25).$$

If the axes of co-ordinates have the special directions adopted in Art. 113, the conditions ξ, η, $\zeta = 0$ give us at once u, v, w in terms of p, q, r, viz.

$$u = -\frac{Lp + L'q + L''r}{A}, \text{ &c., &c. } \dots\dots\dots(26).$$

Substituting these values in the expressions for λ, μ, ν obtained from (23), we find

$$\lambda = \frac{d\Theta}{dp}, \quad \mu = \frac{d\Theta}{dq}, \quad \nu = \frac{d\Theta}{dr} \dots\dots\dots\dots(27),$$

where

$$2\Theta = \mathfrak{P}p^2 + \mathfrak{Q}q^2 + \mathfrak{R}r^2 + 2\mathfrak{P}'qr + 2\mathfrak{Q}'rp + 2\mathfrak{R}'pq\dots..(28);$$

the coefficients in this expression being determined by the formulæ

$$\mathfrak{P} = P - \frac{L^2}{A} - \frac{M^2}{B} - \frac{N^2}{C},$$

$$\mathfrak{P}' = P' - \frac{L'L''}{A} - \frac{M'M''}{B} - \frac{N'N''}{C},$$

&c.,   &c.

These formulæ hold for any case in which the force-constituent of the impulse is zero. Introducing the conditions (25) for steady motion, we have to determine $p : q : r$ the three equations

$$\begin{aligned}
\mathfrak{P}p + \mathfrak{R}'q + \mathfrak{Q}'r &= kp, \\
\mathfrak{R}'p + \mathfrak{Q}q + \mathfrak{P}'r &= kq, \\
\mathfrak{Q}'p + \mathfrak{P}'q + \mathfrak{R}r &= kr
\end{aligned} \right\} \dots\dots\dots\dots(29).$$

The form of (29) shews that the line whose direction-ratios are $p : q : r$ is parallel to one of the principal axes of the ellipsoid

$$\Theta(x, y, z) = \text{const.} \dots\dots\dots\dots(30).$$

There are therefore three permanent screw-motions such that the corresponding impulsive wrench in each case reduces to a couple only. The axes of these three screws are mutually at right angles, but do not in general intersect.

115. We will now shew that in all cases where the impulse consists of a couple only, the motion can be completely determined. It is convenient, retaining the same directions of the axes as before, to change the position of the origin. To transfer the origin to any point (x, y, z) we must write

$$u + ry - qz, \quad v + pz - rx, \quad w + qx - py$$

for u, v, w, respectively. We have then in the expression for the kinetic energy

$$\text{new } M'' = - Bx + M'', \quad \text{new } N' = Cx + N', \&c.,$$

so that if we make

$$2x = \frac{M''}{B} - \frac{N'}{C}, \qquad 2y = \frac{N}{C} - \frac{L''}{A}, \qquad 2z = \frac{L'}{A} - \frac{M}{B} \dots\dots(31),$$

we have, in the new expression for $2T$,

$$\frac{M''}{B} = \frac{N'}{C}, \qquad \frac{N}{C} = \frac{L''}{A}, \qquad \frac{L'}{A} = \frac{M}{B}.$$

Let us denote the values of these pairs of equal quantities by α, β, γ respectively. The formulæ (26) may then be written

$$u = -\frac{d\Psi}{dp}, \qquad v = -\frac{d\Psi}{dq}, \qquad w = -\frac{d\Psi}{dr} \dots\dots\dots\dots(32),$$

where

$$\Psi = \frac{L}{A}p^2 + \frac{M'}{B}q^2 + \frac{N''}{C}r^2 + 2\alpha qr + 2\beta rp + 2\gamma pq\dots\dots(33).$$

The motion of the body at any instant may be conceived as made up of two parts;—a motion of translation equal to that of the origin, and one of rotation about an instantaneous axis passing through the origin. The latter part is to be determined by the equations

$$\frac{d\lambda}{dt} = r\mu - q\nu, \quad \&c., \quad \&c.,$$

or

$$\frac{d}{dt}\frac{d\Theta}{dp} = r\frac{d\Theta}{dq} - q\frac{d\Theta}{dr}, \quad \&c., \quad \&c.$$

L.   9

These are identical in form with the equations of motion of a rigid body about a fixed point, so that we may make use of Poinsot's well-known solution of the latter problem. The angular motion of the body is therefore obtained by making the ellipsoid (30), which is fixed in the body, roll on the plane

$$\lambda x + \mu y + \nu z = \text{const.},$$

which is fixed in space, with an angular velocity proportional to the length OI of the radius vector drawn from the origin to the point of contact I. The representation of the actual motion is then completed by impressing on the whole system of rolling ellipsoid and plane a velocity whose components are given by (32). The direction of this velocity is that of the normal OM to the tangent plane to the quadric

$$\Psi (x, y, z) = - \epsilon^3 \dots\dots\dots\dots\dots\dots\dots\dots\dots(34),$$

at the point P where OI meets this quadric, and its magnitude is

$$\frac{\epsilon^8}{OP . OM} \times \text{angular velocity of body}\dots\dots\dots(35).$$

If OI do not meet (34), but the conjugate quadric obtained by changing the sign of ϵ, the sense of the velocity (35) is reversed.

116.   Of course for particular varieties of the moving solid the expression for $2T$ becomes greatly simplified. For instance:

(a)   let us suppose that the body has a plane of symmetry as regards both its form and the distribution of matter in its interior, and let this plane be taken as that of xy. It is plain that the energy of the motion is unaltered if we reverse the signs of w, p, q, the motion being exactly similar in the two cases. This requires that A', B', P', Q', L, M, L', M', N'' should vanish. One of the directions of permanent translation is then parallel to z. The three screws of Art. 114 are now pure rotations; the axis of one of them is parallel to z; those of the other two are at right angles in the plane xy, but do not in general intersect the first.

(b)   If the body have a second plane of symmetry, at right angles to the former one, let this be taken as the plane of zx. We find in the same way that in this case the coefficients

C', R', N, L'' also must vanish, so that the expression for $2T$ assumes the form

$$2T = Au^2 + Bv^2 + Cw^2$$
$$+ Pp^2 + Qq^2 + Rr^2$$
$$+ 2N'wq + 2M''vr \dots\dots\dots\dots\dots\dots(36).$$

The directions of permanent translation are parallel to the three axes of co-ordinates. The axis of x is the axis of one of the permanent screws (now pure rotations) of Art. 114; and those of the other two intersect it at right angles (being parallel to y and z respectively), though not necessarily in the same point.

(c) If, further, the body be one of revolution, about x, say, the value of $2T$ given by (35) must be unaltered when we write v, q, $-w$, $-r$ for w, r, v, q, respectively; for this is merely equivalent to turning the axes of y, z through a right angle. Hence we must have $B = C$, $Q = R$, $M'' = -N'$. If we further transfer the origin to the point O of Art. 115 we have $M' = N'$. These conditions can be satisfied only by $M'' = 0$, $N' = 0$, so that

$$2T = Au^2 + B(v^2 + w^2)$$
$$+ Pp^2 + Q(q^2 + r^2) \dots\dots\dots\dots\dots(37).$$

(d) If in (b) the body have a third plane of symmetry at right angles to the two former ones, then taking this plane as that of yz we have, evidently,

$$2T = Au^2 + Bv^2 + Cw^2$$
$$+ Pp^2 + Qq^2 + Rr^2 \dots\dots\dots\dots\dots(38).$$

The axes of co-ordinates are in the directions of the three permanent translations; they are also the axes of the three permanent screw-motions (now pure rotations) of Art. 114.

(e) Next let us consider another class of cases. Let us suppose that the body has a sort of skew symmetry about a certain axis (say that of x), viz. that it is identical with itself turned through two right angles about this axis, but has no plane of symmetry*. The expression for $2T$ must be unaltered when we

---

* A two-bladed screw-propeller of a ship is an example of a body of this kind.

change the signs of v, w, q, r, so that the coefficients B', C', Q', R', M, N, L', L'' must all vanish.   We have then

$$2T = Au^2 + Bv^2 + Cw^2 + 2A'vw$$
$$+ Pp^2 + Qq^2 + Rr^2 + 2P'qr$$
$$+ 2Lpu$$
$$+ 2q(M'v + N'w)$$
$$+ 2r(M''v + N''w)\dots\dots\dots\dots(39).$$

The axis of x is one of the directions of permanent translation; and also the axis of one of the three screws of Art. 114, the pitch being $-\dfrac{L}{A}$.   The axes of the two remaining screws intersect it at right angles, but not in general in the same point.

(f) If, further, the body be identical with itself turned through *one* right angle about the above axis*, the expression (39) must be unaltered when v, q, $-w$, $-r$ are written for w, r, v, q, respectively.   This requires that $B = C$, $A' = 0$, $Q = R$, $P' = 0$, $M' = N''$, $N' = -M''$.   If we further transfer the origin to the point chosen in Art. 115 we must have $N' = M''$, and therefore $N' = 0$, $M'' = 0$.   Hence (39) becomes

$$2T = Au^2 + B(v^2 + w^2)$$
$$+ Pp^2 + Q(q^2 + r^2)$$
$$+ 2Lpu$$
$$+ 2M'(vq + wr)\dots\dots\dots\dots(40).$$

(g) If the body possess the same properties of skew symmetry about an axis intersecting the former one at right angles, we evidently must have

$$2T = A(u^2 + v^2 + w^2)$$
$$+ P(p^2 + q^2 + r^2)$$
$$+ 2L(pu + qv + rw)\dots\dots\dots\dots(41).$$

Any direction is now one of permanent translation, and any line drawn through the origin is the axis of a screw of the kind considered in Art. 114, of pitch $-\dfrac{L}{A}$.   The form of (41) is unaltered by any change in the directions of the axes of co-ordinates.

---

* Some four-bladed screw-propellers are examples of bodies of such forms.

117.   In the case (c) of a solid of revolution, the complete determination of the motion (when no external forces act) has been shewn by Kirchhoff[*] to be reducible to a matter of quadratures.

The particular case where the solid moves without rotation about its axis of symmetry, and with this axis always in one plane (*i.e.* when $p = 0$, $q = 0$), has been examined at length by Thomson[†] and Kirchhoff[‡]. The equations (24) then become

$$A\frac{du}{dt} = rBv, \qquad B\frac{dv}{dt} = -rAu,$$
$$R\frac{dr}{dt} = (A - B)uv. \qquad \Big\}\quad \ldots\ldots\ldots\ldots(42).$$

Let X, Y be the co-ordinates at any instant of the moving origin relatively to axes fixed in space in the plane xy, the direction of X being that of the resultant impulse I of the motion; and let θ denote the angle (measured in the positive direction) which x makes with X. We have then

$$Au = I\cos\theta, \quad Bv = -I\sin\theta, \quad r = \dot\theta.$$

The first two of equations (42), which merely express the fixity of the direction of the impulse in space, are satisfied identically; the third gives

$$R\ddot\theta + \frac{A - B}{AB}I^2\sin\theta\cos\theta = 0,$$

or, writing $2\theta = \vartheta$,

$$\ddot\vartheta + \frac{(A - B)I^2}{ABR}\sin\vartheta = 0\ldots\ldots\ldots\ldots\ldots(43),$$

the equation of motion of a common pendulum. When ϑ has been determined so as to satisfy (43) and the initial conditions, X and Y are to be found from the equations

$$\dot X = u\cos\theta - v\sin\theta = \tfrac12 I\left(\frac1A + \frac1B\right) + \tfrac12 I\left(\frac1A - \frac1B\right)\cos\vartheta,$$
$$\dot Y = u\sin\theta + v\cos\theta = \tfrac12 I\left(\frac1A - \frac1B\right)\sin\vartheta = \frac{R}{I}\ddot\vartheta, \quad \Big\}\quad\ldots(44),$$

the second of which gives

$$Y = \frac{R}{I}\dot\vartheta,$$

---

[*] *Crelle*, t. 71. Ueber die Bewegung eines Rotationkörpers in einer Flüssigkeit.
[†] Thomson and Tait, *Natural Philosophy*, Art. 332.
[‡] *l. c.*

the additive constant being zero if the axis of X be taken coincident with, and not merely parallel to, the axis of the impulse I.

The exact solution of (43) involves the use of elliptic functions. The nature of the motion, in the various cases that may arise, is however readily seen from the theory of the simple pendulum. For a full discussion of it we refer to Thomson and Tait, Arts. 333, *et seq.*

It appears from (43) that the motion of the solid parallel to its axis is stable or unstable according as $A \lessgtr B$. Since A denotes twice the kinetic energy of the solid moving with unit velocity parallel to its axis, and similarly for B, it is tolerably obvious that if the solid resemble a prolate ellipsoid of revolution $A < B$, whilst the reverse is the case if it resemble an oblate ellipsoid. Compare Art. 111.

The above analysis applies equally well to the somewhat more general case (*b*) of a body with two mutually perpendicular planes of symmetry, when the motion is altogether parallel to one of these planes. If this plane be that of xy we must suppose the origin transferred to the point $\left(\dfrac{M''}{B}, 0, 0\right)$; if it be that of xz, to the point $\left(-\dfrac{N'}{C}, 0, 0\right)$.

118.   The question of the stability of the motion of a body moving parallel to an axis of symmetry is more simply treated by approximate methods. Thus, in the case (*d*) of a body with three planes of symmetry, and slightly disturbed from a state of steady motion parallel to x, we have, writing $u = c + u'$, and assuming u', v, w, p, q, r to be all small,

$$A\frac{du'}{dt} = 0, \quad B\frac{dv}{dt} = -Acr, \qquad C\frac{dw}{dt} = Acq,$$

$$P\frac{dp}{dt} = 0, \quad Q\frac{dq}{dt} = (C - A)cw, \quad R\frac{dr}{dt} = (A - B)cv.$$

Hence

$$B\frac{d^2v}{dt^2} + A\frac{(A - B)}{R}c^2v = 0,$$

with a similar equation for r, and

$$C\frac{d^2w}{dt^2} + A\frac{(A-C)}{R}c^2 w = 0,$$

with a similar equation for q. The motion is therefore stable only if A be greater than either B or C. It appears from Art. 111 that the only direction of stable motion of an ellipsoid is that of its least axis. For practical illustrations of this result see Thomson and Tait, Art. 336.

119. If in (24) we write $T = \mathfrak{T} + \mathfrak{T}'$, and separate the terms due to $\mathfrak{T}$ and $\mathfrak{T}'$ respectively, we obtain expressions for the forces exerted on the moving solid by the pressure of the surrounding fluid; viz. we have for the total component ($\mathfrak{X}$, say,) of the fluid pressure parallel to x

$$\mathfrak{X} = -\frac{d}{dt}\frac{d\mathfrak{T}}{du} + r\frac{d\mathfrak{T}}{dv} - q\frac{d\mathfrak{T}}{dw},$$

and for the moment ($\mathfrak{L}$) of the same pressures about x,

$$\mathfrak{L} = -\frac{d}{dt}\frac{d\mathfrak{T}}{dp} + w\frac{d\mathfrak{T}}{dv} - v\frac{d\mathfrak{T}}{dw} + r\frac{d\mathfrak{T}}{dq} - q\frac{d\mathfrak{T}}{dr}.$$

The forms of these expressions being known, it is not difficult to verify them by direct calculation from the formula (18). We should thus obtain an independent though somewhat tedious proof of the general equations of motion (24).

If the body be constrained to move with a uniform velocity of translation, the components of which, relatively to the axes of Art. 113, are u, v, w, we have $\mathfrak{X}$, $\mathfrak{Y}$, $\mathfrak{Z} = 0$, so that the effect of the fluid pressure is represented by a *couple* whose components are

$$\mathfrak{L} = (B-C)vw, \quad \mathfrak{M} = (C-A)wu, \quad \mathfrak{N} = (A-B)uv\ldots(45).$$

The coefficients A, B, C in the expression for $2T$ differ from those in the expression for $2\mathfrak{T}$ only by the addition of the mass of the solid, so that it is immaterial in (21) which set of coefficients we understand by these symbols.

If we draw in the ellipsoid

$$Ax^2 + By^2 + Cz^2 = \text{const} \ldots \ldots \ldots (46),$$

a radius-vector r in the direction of the velocity (u, v, w) and erect the perpendicular h from the centre to the tangent plane at the extremity of r, the plane of the above couple is that of h and r, and its magnitude is proportional to $\sin \hat{hr}$ directly, and to h inversely. Its tendency is to turn the body from r to h. Let us suppose that A, B, C are in order of magnitude, and that the direction of the velocity (u, v, w) deviates but slightly from that of one of the principal axes of (46). If this axis be that of x, the tendency of the above couple is to diminish, and if that of z, to increase the deviation; whilst in the case of a slight deviation from the axis of y the tendency of the couple depends on the position of r relative to the principal circular sections of (46). Compare Art. 118.

Case of a Perforated Solid.

120.   If the moving solid have one or more apertures or perforations, so that the space external to it is multiply-connected, the fluid may have a motion independent of that of the solid, viz. a cyclic motion in which the circulations in the various non-evanescible circuits which can be drawn through the apertures may have any values whatever. We will briefly indicate how the foregoing methods may be adapted to this case. Let $\kappa_1, \kappa_2 \ldots\ldots$ be the values of the circulations in the above-mentioned circuits, and let $d\sigma_1, d\sigma_2, \ldots$ be surface-elements of the corresponding barriers necessary (as explained in Art. 54) to reduce the region occupied by the fluid to a simply-connected one. Further, let l, m, n denote the direction-cosines of the normal drawn towards the fluid at any point of the surface of the solid, or drawn on the positive side at any point of a barrier. We may now write

$$\phi = u\phi_1 + v\phi_2 + w\phi_3 + p\chi_1 + q\chi_2 + r\chi_3 + \kappa_1\omega_1 + \kappa_2\omega_2 + \ldots(47).$$

The functions ϕ, χ are determined by the same conditions as before. To determine ω_1 we have the conditions

(a)   that it must satisfy $\nabla^2\omega_1 = 0$ throughout the fluid;

(b)   that its derivatives must vanish at infinity;

(c)   that $\dfrac{d\omega_1}{dn} = 0$ at the surface of the solid; and

(d) that ω_1 must be a monocyclic function, the cyclic constant being unity; viz. the increment of the function must be unity when the point to which it refers describes a circuit cutting the first barrier once and once only, and zero when the point describes a circuit not cutting this barrier.

It appears from Art. 62 that these conditions completely determine ω_1, save as to an additive constant.

The energy of motion of the fluid is given by Art. 67, viz. we have

$$2\mathfrak{T} = -\rho \iint \phi \frac{d\phi}{dn} dS + \rho \kappa_1 \iint \frac{d\phi}{dn} d\sigma_1 + \rho \kappa_2 \iint \frac{d\phi}{dn} d\sigma_2 + \dots$$

Substituting the values of ϕ, $\dfrac{d\phi}{dn}$ from (47) we obtain a homogeneous expression of the second degree in u, v, w, ... , κ_1, κ_2, ... This expression consists of three parts. The first is a homogeneous quadratic function of u, v, w, p, q, r, the coefficients in which are given by the same formulæ as in Art. 110; the second part consists of products of u, v, w,... into κ_1, κ_2...; whilst the third part is a quadratic function of the coefficients κ. The coefficients of the second part all vanish. Thus the coefficient of $u\kappa_1$ is

$$-\rho \iint \left(\omega_1 \frac{d\phi_1}{dn} + \phi_1 \frac{d\omega_1}{dn} \right) dS + \rho \iint \frac{d\phi_1}{dn} d\sigma_1,$$

and to see that the value of this expression is in fact zero, we have only to compare (30) and (31) of Art. 66, writing $\phi = \phi_1$, $\psi = \omega_1$, and therefore $\kappa_1 = \kappa_2 = \dots = 0$, $\kappa_1' = 1$, $\kappa_2' = \kappa_3' = \dots = 0$. The coefficients of the third part are found as follows. We have

$$\text{coeff. of } \kappa_1{}^2 = -\rho \iint \omega_1 \frac{d\omega_1}{dn} dS + \rho \iint \frac{d\omega_1}{dn} d\sigma_1 = \rho \iint \frac{d\omega_1}{dn} d\sigma_1,$$

$$\text{coeff. of } \kappa_1\kappa_2 = -\rho \iint \left(\omega_1 \frac{d\omega_2}{dn} + \omega_2 \frac{d\omega_1}{dn} \right) dS + \rho \iint \frac{d\omega_2}{dn} d\sigma_1 + \rho \iint \frac{d\omega_1}{dn} d\sigma_2$$

$$= 2\rho \iint \frac{d\omega_2}{dn} d\sigma_1 = 2\rho \iint \frac{d\omega_1}{dn} d\sigma_2,$$

by another simple application of Thomson's extension of Green's theorem.

Hence the total energy is obtained by adding to the right-hand side of (21) an expression of the form

$$\Sigma K_{rr} \kappa_r^2 + 2\Sigma K_{rs} \kappa_r \kappa_s,$$

where

$$K_{rr} = \rho \iint \frac{d\omega_r}{dn} d\sigma_r,$$

$$K_{rs} = \rho \iint \frac{d\omega_r}{dn} d\sigma_s = \rho \iint \frac{d\omega_s}{dn} d\sigma_r.$$

121. The impulsive forces necessary to produce from rest the actual motion at any instant now consist partly of impulsive forces applied to the solid, and partly (as explained in Art. 61) of impulsive pressures $\rho\kappa_1$, $\rho\kappa_2$, &c. uniform over the several membranes which are supposed for a moment to occupy the positions of the barriers above-mentioned. The components of the force- and couple-resultants of the first set, we denote by ξ_1, η_1, ζ_1, and λ_1, μ_1, ν_1, respectively; those of the force and couple equivalent to the second set by ξ_2, η_2, ζ_2, and λ_2, μ_2, ν_2. By the 'impulse' of the motion at any instant we shall understand the force and couple equivalent to both these sets combined, so that if ξ, η, ζ; λ, μ, ν be its components, we have

$$\xi = \xi_1 + \xi_2, \text{ &c., &c.,}$$
$$\lambda = \lambda_1 + \lambda_2, \text{ &c., &c.}$$

If we use the term 'impulse' in this sense, the reasoning of Art. 108 and consequently the equations of motion (20) will still hold. The formulæ (23), however, connecting ξ, η, ζ, &c. with T require correction.

By the same reasoning, and with the same notation as in Art. 112, we have

$$\frac{d\mathbb{T}'}{du} = \xi_1 + \rho \iint \phi l dS$$

$$= \xi_1 + \rho \iint (u\phi_1 + \dots + p\chi_1 + \dots + \kappa_1\omega_1 + \dots) \frac{d\phi_1}{dn} dS$$

$$= \xi_1 - \frac{d\mathbb{T}}{du} + \rho\kappa_1 \iint \omega_1 \frac{d\phi_1}{dn} dS + \text{&c.}$$

$$\frac{d\mathfrak{T}'}{dp} = \lambda_1 + \rho \iint \phi\,(ny - mz)\,dS$$

$$= \lambda_1 + \rho \iint (u\phi_1 + \ldots + \kappa_1\omega_1 + \ldots)\,\frac{d\chi_1}{dn}\,dS$$

$$= \lambda_1 - \frac{d\mathfrak{T}}{dp} + \rho\kappa_1 \iint \omega_1 \frac{d\chi_1}{dn}\,dS + \&\mathrm{c.};$$

so that

$$\xi_1 = \frac{dT}{du} - \rho\kappa_1 \iint \omega_1 \frac{d\phi_1}{dn}\,dS - \&\mathrm{c.},$$

$$\lambda_1 = \frac{dT}{dp} - \rho\kappa_1 \iint \omega_1 \frac{d\chi_1}{dn}\,dS - \&\mathrm{c.}$$

We saw above that

$$\iint \omega_1 \frac{d\phi_1}{dn}\,dS = \iint \frac{d\phi_1}{dn}\,d\sigma_1, \qquad \iint \omega_1 \frac{d\chi_1}{dn}\,dS = \iint \frac{d\chi_1}{dn}\,d\sigma_1, \ \&\mathrm{c.},$$

so that we may also write

$$\xi_1 = \frac{dT}{du} - \rho\kappa_1 \iint \frac{d\phi_1}{dn}\,d\sigma_1 - \&\mathrm{c.},$$

$$\lambda_1 = \frac{dT}{dp} - \rho\kappa_1 \iint \frac{d\chi_1}{dn}\,d\sigma_1 - \&\mathrm{c.}$$

Again, by Statics,

$$\xi_2 = \rho\kappa_1 \iint l\,d\sigma_1 + \&\mathrm{c.},$$

$$\lambda_2 = \rho\kappa_1 \iint (ny - mz)\,d\sigma_1 + \&\mathrm{c.},$$

whence finally,

$$\xi = \xi_1 + \xi_2 = \frac{dT}{du} + \xi_0,$$

$$\lambda = \lambda_1 + \lambda_2 = \frac{dT}{dp} + \lambda_0,$$

with similar expressions for the remaining components of the impulse. We have here written for shortness

$$\xi_0 = \rho\kappa_1 \iint \left(l - \frac{d\phi_1}{dn}\right) d\sigma_1 + \&\mathrm{c.},$$

$$\lambda_0 = \rho\kappa_1 \iint \left(ny - mz - \frac{d\chi_1}{dn}\right) d\sigma_1 + \&\mathrm{c.}$$

$$\&\mathrm{c.} \qquad\qquad \&\mathrm{c.}$$

It is plain that ξ_0, η_0 ζ_0, λ_0, μ_0, ν_0, are the components of the impulse of the cyclic fluid motion which remains when the solid is (by forces applied to it alone) brought to rest.

122.   As a simple example we may take the case treated by Thomson*; viz. where the solid is a circular ring (of any form of section), and has therefore only one aperture.   If we take the axis of the ring as axis of x, we see by the same reasoning as in Art. 116 that if the situation of the origin in this axis be properly chosen we may write

$$2T = Au^2 + B(v^2 + w^2)$$
$$+ Pp^2 + Q(q^2 + r^2)$$
$$+ K\kappa^2.$$

Hence $\qquad \xi = Au + \xi_0, \qquad \eta = Bv, \qquad \zeta = Bw,$

$\qquad\qquad\quad \lambda = Pp, \qquad\quad \mu = Qq, \qquad \nu = Rr.$

The fourth of equations (20) then gives $\dfrac{dp}{dt} = 0$, or $p = $ const. as is obviously the case.   Let us suppose that $p = 0$, and that the ring is slightly disturbed from a state of steady motion parallel to its axis.   In the beginning of the disturbed motion v, w, q, r are small quantities whose squares and products we may neglect.   The first of (20) then gives $\dfrac{du}{dt} = 0$, or $u = $ const., and the remaining equations become

$$B\frac{dv}{dt} = -(Au + \xi_0)r, \qquad Q\frac{dq}{dt} = -\{(A - B)u + \xi_0\}w,$$

$$B\frac{dw}{dt} = \ \ (Au + \xi_0)q, \qquad Q\frac{dr}{dt} = \ \ \{(A - B)u + \xi_0\}v.$$

Eliminating r, we find

$$BQ\frac{d^2v}{dt^2} = -(Au + \xi_0)\{(A - B)u + \xi_0\}v \ldots\ldots(48).$$

Exactly the same equation is satisfied by w. It is therefore necessary and sufficient for stability that the coefficient of v on the

* *Phil. Mag.* Nov. 1871.

right-hand side of (48) should be negative; and the time of a small oscillation, in the case of disturbed stable motion, is

$$2\pi \left[\frac{BQ}{(Au + \xi_0)\{(A - B)u + \xi_0\}} \right]^{\frac{1}{2}}.$$

123. The general equations of motion of the ring are also satisfied by ξ, η, ζ, λ, $\mu = 0$, and ν constant. We have then

$$u = -\frac{\xi_0}{A}, \qquad r = \text{const.}$$

The motion of the ring is then one of uniform rotation about an axis in the plane yz parallel to that of y, and at a distance $\dfrac{u}{r}$ from it.

Case of two or more moving solids.

124. The foregoing methods fail when we have two or more moving solids, or when the fluid does not extend in all directions to infinity, being bounded externally by fixed rigid walls. In such cases we may suppose the position at the time t of each moving solid to be defined by means of six 'co-ordinates,' in the manner explained in treatises on Kinematics. It is easy to see that ϕ must be a linear function of the rates of variation of these co-ordinates (in other words, of the 'generalized velocity-components' of the system), and thence that the kinetic energy of the system is, as in Art. 110. a homogeneous quadratic function of these generalized velocities, with however the important change that the coefficients in this function are not constants, but themselves functions of the co-ordinates of the system. The equations of motion are then most conveniently formed by Lagrange's method[*], the applicability of which to systems of the peculiar kind here considered requires however to be in the first place established[†].

The accompanying references will be of service to the reader who wishes to pursue the study of the general problem in the manner indicated. We content ourselves here with the discussion

---

[*] See Thomson and Tait, Art. 329.

[†] See Thomson, *Phil. Mag.* May, 1873, and Kirchhoff, *Vorlesungen über Math. Physik. Mechanik,* c. 19, § 1.

of a very simple case in which the forces acting on the solids can be readily calculated by the direct method.

125.   Let us suppose that we have two spheres in motion in the line joining their centres A, B. Let u be the velocity of the first in the direction AB, v that of the second in the direction BA. Further, P being any point of the fluid, let

$$PA = r, \qquad PB = s,$$
$$\angle PAB = \theta, \qquad \angle PBA = \chi;$$

also let a, b be the radii of the spheres and c the distance AB of their centres.   If the sphere B were absent, and its place occupied by fluid, the velocity-potential ϕ_1 due to the sphere A alone would be, by Art. 102,

$$\phi_1 = -\tfrac{1}{2}\frac{ua^3}{r^2}\cos\theta.$$

To find the value of ϕ_1 in the neighbourhood of B we have

$$r^2 = c^2 - 2cs\cos\chi + s^2,$$
$$r\cos\theta = c - s\cos\chi,$$

so that

$$\phi_1 = -\tfrac{1}{2}\frac{ua^3}{c^2}\left(1 - \frac{s}{c}\cos\chi\right)\left(1 - 2\frac{s}{c}\cos\chi + \frac{s^2}{c^2}\right)^{-\frac{3}{2}}$$
$$= -\tfrac{1}{2}\frac{ua^3}{c^2}\left(1 + 2\frac{s}{c}\cos\chi + 3\frac{s^2}{c^2}\cdot\frac{3\cos^2\chi - 1}{2} + \&c.\right)^{*}.$$

This gives at the surface of B,

$$\frac{d\phi_1}{ds} = -\tfrac{1}{2}\frac{ua^3}{c^3}\left(2\cos\chi + 2.3.\frac{b}{c}.\frac{3\cos^2\chi - 1}{2} + \&c.\right).$$

The relation which actually holds at the surface of B, viz.

$$\frac{d\phi}{ds} = v\cos\chi.$$

* We recognize the coefficients of $2\frac{s}{c}$, $3\frac{s^2}{c^2}$, &c., within the brackets, as the 'zonal harmonics' of orders 1, 2, &c. respectively.   In fact, remembering that $\frac{\cos\theta}{r^2}$ is the potential at the point P due to a small magnet of unit moment placed at A with its axis pointing in the direction AB, we readily find from the definition of the aforesaid zonal harmonics Q_1, Q_2, &c., that

$$\phi_1 = \tfrac{1}{2}ua^3.\frac{d}{dc}\left(\frac{1}{c} + \frac{s}{c^2}Q_1 + \frac{s^2}{c^3}Q_2 + \&c.\right).$$

is therefore satisfied by making

$$\phi = \phi_1 + \phi_2,$$

where

$$\phi_2 = -\tfrac{1}{2}\frac{vb^3}{s^2}\cos\chi - \tfrac{1}{2}\frac{ua^3}{c^3}\left(\frac{b^3}{s^2}\cos\chi + 2\frac{b^5}{cs^3}\cdot\frac{3\cos^2\chi - 1}{2} + \&\text{c.}\right).$$

The condition at the surface of A, viz.

$$\frac{d\phi}{ds} = u\cos\theta,$$

is however no longer satisfied; but it is plain from the course of the above investigation that the error in the normal velocity there will be of the order $\dfrac{a^3b^3}{c^6}$, and that if this be rectified by the addition of a properly chosen term ϕ_3 to the above value of ϕ, the effect of this at the surface of B will be of the order $\dfrac{a^6b^3}{c^9}$. In the particular cases examined below we shall suppose a and b both small in comparison with c, and shall not take into account small quantities of so high an order as that last written. We have then at the surface of B

$$\phi = -\tfrac{1}{2}b\left(v + 3u\frac{a^3}{c^3}\right)\cos\chi - \tfrac{5}{2}\frac{a^3b^2}{c^4}u\cdot\frac{3\cos^2\chi - 1}{2} - \&\text{c.}\ldots(49),$$

$$\frac{d\phi}{ds} = -v\cos\chi,$$

$$\frac{d\phi}{bd\chi} = \tfrac{1}{2}\left(v + 3u\frac{a^3}{c^3}\right)\sin\chi + 1\tfrac{1}{2}\frac{a^3b}{c^4}u\sin\chi\cos\chi + \&\text{c.}$$

The total effect of the fluid pressure on the sphere B evidently reduces to a force in the direction AB, the amount of which is

$$\int_0^\pi p\cdot 2\pi b\sin\chi\cdot bd\chi\cdot\cos\chi\ldots\ldots\ldots\ldots(50),$$

where p is to be found from (18). In calculating $\dfrac{d\phi}{dt}$ we must remember, as in Art. 102, that the origin B of the polar co-ordinates s, χ is itself in motion with velocity v in the direction BA. The rates at which the values of s, χ for a fixed point are increasing in consequence of this motion are easily seen to be $- v \cos \chi$, and $\frac{v}{b} \sin \chi$, respectively, so that we must write for $\frac{d\phi}{dt}$,

$$-\tfrac{1}{2}b \frac{d}{dt}\left(v + \frac{3a^3}{c^3} u\right) \cos \chi + \tfrac{15}{2} \frac{a^3 b}{c^4} uv \sin^2\chi \cos \chi + \&c.,$$

where terms which obviously contribute nothing to the integral (50) have been omitted. Again

$$\tfrac{1}{2}(\text{vel.})^2 = \tfrac{1}{2}\left(\frac{d\phi}{ds}\right)^2 + \tfrac{1}{2}\left(\frac{d\phi}{bd\chi}\right)^2$$

$$= \ldots + \tfrac{15}{4}\frac{a^3 b}{c^4} uv \sin^2\chi \cos \chi + \&c.\ldots\ldots\ldots(51),$$

similar omissions being made. Now

$$\int_0^\pi \sin \chi \cos^2\chi\, d\chi = \tfrac{2}{3}, \qquad \int_0^\pi \sin^3\chi \cos^2\chi\, d\chi = \tfrac{4}{15},$$

so that we have finally for the resultant fluid pressure on B in the direction AB,

$$\tfrac{2}{3}\pi\rho b^3 \frac{d}{dt}\left(v + \frac{3a^3}{c^3} u\right) - \frac{6\pi\rho a^3 b^3}{c^4} uv \ \ldots\ldots(52).$$

This result is correct to the order $\frac{a^3 b^3}{c^4}$ inclusive. Since

$$\frac{dc}{dt} = -(u + v),$$

(52) may also be written

$$\tfrac{2}{3}\pi\rho b^3 \left(\frac{dv}{dt} + \frac{3a^3}{c^3}\frac{du}{dt}\right) + \frac{6\pi\rho a^3 b^3}{c^4} u^2 \ \ldots\ldots\ldots(53).$$

We proceed to examine some particular cases, keeping only the most important terms in each.

(a)   Let $b = a$, $v = u$, so that the motion is symmetrical with respect to the plane bisecting AB at right angles, and is the same as if this plane formed a rigid boundary to the fluid on either side of it. We have thus the solution of the case where a sphere moves directly towards or away from a fixed plane wall. The force repelling the sphere from the wall is *

$$\tfrac{1}{2}M'\left(1 + \frac{3a^3}{c^3}\right)\frac{du}{dt} + \&c.,$$

* Stokes, *Camb. Trans.* Vol. VIII. (1843).

where M' is the mass of fluid displaced by the sphere. Hence the principal effect of the plane boundary is to increase the inertia of the sphere in the ratio $1 + \dfrac{3a^3}{c^3} : 1$, c denoting double the distance of the centre of the sphere from the plane.

(b) Let us suppose each sphere constrained to move with constant velocity. The force which must be applied to B in order to maintain this motion is $\dfrac{6\pi\rho a^3 b^3}{c^4} u^2$ approximately, and is in the direction BA. The spheres therefore appear to repel one another. The forces to be applied to the two spheres are not equal and opposite except when $v = u$.

(c) Let us suppose that each sphere makes small periodic oscillations about a mean position, the period being the same for each. The average value of the first term of (52) is then zero, and the mutual action of the two spheres is equivalent to a force $\dfrac{6\pi\rho a^3 b^3}{c^4} \overline{uv}$, urging them together, where $\overline{uv}$ denotes the mean value of uv. If u, v differ in phase by less than a quarter-period this force is one of attraction, if by more than a quarter-period it is one of repulsion.

(d) Let A perform small periodic oscillations while B is held at rest. The mean force on B is now zero to our order of approximation. To carry the approximation further, we remark that the mean value of $\dfrac{d\phi}{dt}$ at the surface of B is necessarily zero, and that the next important term in the value (51) of the semi-square of the velocity is, when $v = 0$, $\frac{45}{4} \dfrac{a^6 b}{c^7} u^2 \sin^2\chi \cos\chi$, and the resulting term in (50) is found on integration to be $\dfrac{6\pi\rho a^6 b^3}{c^7} \bar{u}^2$, where $\bar{u}^2$ denotes the average value of the square of the velocity of A.

This result comes under a general principle enunciated by Thomson. If we have two bodies immersed in a fluid, one of which A performs small vibrations while the other B is held at rest, the fluid velocity at the surface of B will on the whole be greater on the side nearer A than on that which is more remote. Hence by (18) the average* pressure on the former side will be less than that on the latter, so that B will experience on the whole an attraction towards A. As practical illustrations of this principle we may cite the apparent attraction of a delicately-suspended card by a vibrating tuning-fork, and other similar phenomena studied experimentally by Guthrie† and explained in the above manner by Thomson§.

The same principle accounts for the indraught of a light powder, strewn on a vibrating plate, towards the ventral segments.

---

* Since ϕ is by hypothesis a periodic function of t, the term $\frac{d\phi}{dt}$ in (18) contributes nothing to the *average* effect.

† *Proc. R. S.*

§ *Reprint*, Art. XLI.

CHAPTER VI.

126. So far our investigations have been confined for the most part to the case of irrotational motion. We now proceed to the study of rotational or 'vortex' motion. This subject was first investigated by Helmholtz, in Crelle's Journal, 1858; other and simpler proofs of some of his theorems were afterwards given by Thomson in the paper on vortex motion already cited in Chapter III.

A line drawn from point to point so that its direction is everywhere that of the instantaneous axis of rotation of the fluid is called a 'vortex-line.' The differential equations of the system of vortex-lines are

$$\frac{dx}{\xi} = \frac{dy}{\eta} = \frac{dz}{\zeta},$$

where ξ, η, ζ have, as throughout this chapter, the meanings assigned in Art. 38.

If through every point of a small closed curve we draw the corresponding vortex-line, we obtain a tube, which we call a 'vortex-tube.' The fluid contained within such a tube constitutes what is called a 'vortex-filament,' or simply a 'vortex.'

Kinematical Theorems.

127. Let ABC, $A'B'C'$ be any two circuits drawn on the surface of a vortex-tube and embracing it, and let AA' be a

connecting line also drawn on the surface. Let us apply the theorem of Art. 40 to the circuit $ABCAA'C'B'A'A$ and the part

Fig. 10.

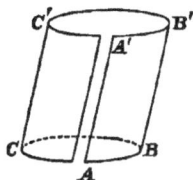

of the surface of the tube bounded by it. Since $l\xi + m\eta + n\zeta$ is zero at every point of this surface, the line-integral

$$\int (udx + vdy + wdz),$$

taken round the circuit, must vanish; *i.e.* in the notation of Art. 39

$$I(ABCA) + I(AA') + I(A'C'B'A') + I(A'A) = 0,$$

which reduces to

$$I(ABCA) = I(A'B'C'A').$$

Hence the circulation is the same in all circuits embracing the same vortex-tube.

Again, it appears from Art. 39 that the circulation round the boundary of any cross-section of the tube, made normal to its length, is $2\omega\sigma$, where $\omega = (\xi^2 + \eta^2 + \zeta^2)^{\frac{1}{2}}$ is the angular velocity of the fluid at the section, and σ the (infinitely small) area of the section.

Combining these results we see that the product of the angular velocity into the cross-section is the same at all points of a vortex. This product is conveniently termed the 'strength' of the vortex.

The foregoing proof is due to Thomson; the theorem itself was first given by Helmholtz, who deduced it from the relation

$$\frac{d\xi}{dx} + \frac{d\eta}{dy} + \frac{d\zeta}{dz} = 0 \dots\dots\dots\dots\dots(1),$$

which follows at once from the values of ξ, η, ζ given in Art. 38. In fact, writing in Art. 64, *Cor.* 1, ξ, η, ζ for u, v, w, respectively, we find

$$\iint (l\xi + m\eta + n\zeta)\, dS = 0 \dots\dots\dots\dots\dots(2),$$

where the integration extends over any closed surface lying wholly in the fluid. Applying this to the closed surface formed by two cross-sections of a vortex-tube, and the portion of the tube intercepted between them, we find $\omega_1 \sigma_1 = \omega_2 \sigma_2$, where ω_1, ω_2 denote the angular velocities at the sections σ_1, σ_2, respectively.

Thomson's proof shews that the theorem is true even when ξ, η, ζ are discontinuous (in which case there may be an abrupt bend at some point of a vortex), provided only that u, v, w are continuous.

An important consequence of the above theorem is that a vortex-line cannot begin or end at any point in the interior of the fluid. Any vortex-lines which exist must either form closed curves, or else traverse the fluid, beginning and ending on its boundaries. Compare Art. 44.

The theorem (6) of Art. 40 may now be enunciated as follows: The circulation in any circuit is equal to twice the sum of the strengths of all the vortices which it embraces.

128. The motion of the fluid occupying any simply-connected region is determinate when we know the values of the expansion (θ, say,) and of the component angular velocities ξ, η, ζ at every point of the region, and the value of the normal velocity (λ, say,) at every point of the boundary.

If possible, let there be two sets of values, u_1, v_1, w_1, and u_2, v_2, w_2, of the component velocities, each satisfying the above conditions; viz. each set satisfying the differential equations

$$\frac{du}{dx} + \frac{dv}{dy} + \frac{dw}{dz} = \theta \dots \dots \dots \dots (3),$$

$$\frac{dw}{dy} - \frac{dv}{dz} = 2\xi, \quad \frac{du}{dz} - \frac{dw}{dx} = 2\eta, \quad \frac{dv}{dx} - \frac{du}{dy} = 2\zeta \dots \dots (4),$$

throughout the region, and the condition

$$lu + mv + nw = \lambda \dots \dots \dots \dots \dots (5),$$

at the boundary. Hence the quantities

$$u' = u_1 - u_2, \qquad v' = v_1 - v_2, \qquad w' = w_1 - w_2,$$

will satisfy (3), (4), and (5) with θ, ξ, η, ζ, λ put each $= 0$; that is to say u', v', w' are the components of the irrotational motion of an incompressible fluid occupying a simply-connected region whose boundary is at rest. Hence (Art. 47) these quantities all vanish; and there is only one possible motion satisfying the given conditions.

The above theorem—an extension of one given in Art. 49—is equally true when the region extends to infinity, and (5) is replaced by the condition that the fluid is there at rest.

129. If, in the last-mentioned case, all the vortices present are within a finite distance of the origin, the complete determination of u, v, w in terms of θ, ξ, η, ζ can be effected, as follows*.

Let us assume

$$u = \frac{dP}{dx} + \frac{dN}{dy} - \frac{dM}{dz},$$
$$v = \frac{dP}{dy} + \frac{dL}{dz} - \frac{dN}{dx}, \quad \right\} \quad \ldots\ldots\ldots\ldots\ldots(6),$$
$$w = \frac{dP}{dz} + \frac{dM}{dx} - \frac{dL}{dy},$$

and seek to determine P, L, M, N so as to satisfy (3) and (4) and make u, v, w zero at infinity. We must have in the first place

$$\nabla^2 P = \theta \ldots\ldots\ldots\ldots\ldots\ldots\ldots(7).$$

Again,

$$2\xi = \frac{dw}{dy} - \frac{dv}{dz} = \frac{d}{dx}\left(\frac{dL}{dx} + \frac{dM}{dy} + \frac{dN}{dz}\right) - \nabla^2 L.$$

Hence, provided

$$\frac{dL}{dx} + \frac{dM}{dy} + \frac{dN}{dz} = 0 \ldots\ldots\ldots\ldots\ldots(8),$$

we have

$$\nabla^2 L = -2\xi, \quad \nabla^2 M = -2\eta, \quad \nabla^2 N = -2\zeta \ldots\ldots(9).$$

Now (7) and (9) are satisfied by making P, L, M, N equal to the potentials of distributions of matter whose densities at the point (x, y, z) are $-\dfrac{\theta}{4\pi}$, $\dfrac{\xi}{2\pi}$, $\dfrac{\eta}{2\pi}$, $\dfrac{\zeta}{2\pi}$, respectively. This gives

---

* See Stokes, *Camb. Trans.* Vol. IX. (1849), and Helmholtz, *Crelle*, t. LV. (1858).

$$P = -\frac{1}{4\pi} \iiint \frac{\theta'}{r} \, dx' \, dy' \, dz' \dots \dots \dots \dots (10),$$

$$\left.
\begin{aligned}
L &= \frac{1}{2\pi} \iiint \frac{\xi'}{r} \, dx' \, dy' \, dz', \\
M &= \frac{1}{2\pi} \iiint \frac{\eta'}{r} \, dx' \, dy' \, dz', \\
N &= \frac{1}{2\pi} \iiint \frac{\zeta'}{r} \, dx' \, dy' \, dz'
\end{aligned}
\right\} \dots \dots \dots \dots \dots (11),$$

where the accents attached to θ, ξ, η, ζ denote the values of these quantities at the point (x', y', z') and r stands for the distance

$$\{(x - x')^2 + (y - y')^2 + (z - z')^2\}^{\frac{1}{2}}.$$

The integrations are supposed to include all parts of space at which θ, ξ, η, ζ have values different from zero.

We must now examine whether the above values of L, M, N really satisfy (8). Since $\frac{d}{dx} \frac{1}{r} = -\frac{d}{dx'} \frac{1}{r}$, &c., &c., we have

$$\frac{dL}{dx} + \frac{dM}{dy} + \frac{dN}{dz}$$

$$= -\frac{1}{2\pi} \iiint \left(\xi' \frac{d}{dx'} \frac{1}{r} + \eta' \frac{d}{dy'} \frac{1}{r} + \zeta' \frac{d}{dz'} \frac{1}{r} \right) dx' \, dy' \, dz'$$

$$= -\frac{1}{2\pi} \iint \frac{l\xi' + m\eta' + n\zeta'}{r} \, dS'$$

$$+ \frac{1}{2\pi} \iiint \left(\frac{d\xi'}{dx'} + \frac{d\eta'}{dy'} + \frac{d\zeta'}{dz'} \right) \frac{dx' \, dy' \, dz'}{r} \dots \dots \dots \dots (12),$$

by (17), Art. 64. The volume-integral vanishes by (1), and the surface-integral vanishes because by hypothesis we have ξ, η, $\zeta = 0$ at all points of the (infinite) surface over which it is taken. Hence (8) is satisfied, and the values (6) of u, v, w satisfy (3) and (4). They also evidently vanish at infinity.

The above results hold even when θ, ξ, η, ζ are discontinuous functions, provided only that u, v, w be continuous. As regards θ this is obvious; but a discontinuity in ξ, η, ζ will necessitate a modification in (12). Let us suppose that as we cross a certain surface Σ the values of ξ, η, ζ change abruptly, and let us distinguish the value on the two sides by suffixes. Two cases present themselves; the vortex-lines may be tangential to Σ on both sides, or they may cross the surface, experiencing there an abrupt change of direction. In the first case we have

$$l\xi_1 + m\eta_1 + n\zeta_1 = l\xi_2 + m\eta_2 + n\zeta_2 = 0 \dots\dots\dots(13)$$

at Σ; and in the second we have

$$l\xi_1 + m\eta_1 + n\zeta_1 = l\xi_2 + m\eta_2 + n\zeta_2\dots\dots(14).$$

In fact, if $d\Sigma$ be a section of a vortex, taken parallel and infinitely close to Σ on one side of it, the product $(l\xi_1 + m\eta_1 + n\zeta_1)\,d\Sigma$ measures the strength of the vortex, which is (Art. 127) the same on both sides of Σ. Now in (12) the region through which the triple integration extends is divided by the surfaces Σ into a certain number of distinct portions. For each of these, taken by itself, the equality of the second and third members of (12) holds; and if we add the results thus obtained, we see that to make (10) true for the region taken as a whole we must add to the third member terms of the form

$$-\frac{1}{2\pi} \iint \{l(\xi_1 - \xi_2) + m(\eta_1 - \eta_2) + n(\zeta_1 - \zeta_2)\} \frac{d\Sigma}{r},$$

due to the two sides of each of the surfaces Σ. The relations (13) and (14) shew however that these terms all vanish, so that (8) is still satisfied.

130. Let us examine the result obtained in Art. 129; and let us suppose first that the fluid is incompressible, so that $\theta = 0$, and therefore $P = 0$. Denoting by $\delta u, \delta v, \delta w$ the portions of u, v, w arising from the element $dx'dy'dz'$ in the integrals (11), we find

$$\delta u = \frac{1}{2\pi} \left(\zeta' \frac{d}{dy} \frac{1}{r} - \eta' \frac{d}{dz} \frac{1}{r} \right) dx'dy'dz',$$

or

$$\delta u = \frac{1}{2\pi r^2} \left(\eta' \frac{z - z'}{r} - \zeta' \frac{y - y'}{r} \right) dx'dy'dz',$$

and similarly

$$\delta v = \frac{1}{2\pi r^2} \left(\zeta' \frac{x - x'}{r} - \xi' \frac{z - z'}{r} \right) dx'dy'dz', \left.\begin{array}{c} \\ \\ \\ \end{array}\right\} \dots\dots(15).$$

$$\delta w = \frac{1}{2\pi r^2} \left(\xi' \frac{y - y'}{r} - \eta' \frac{x - x'}{r} \right) dx'dy'dz'.$$

It appears from the form of these expressions that the resultant of δu, δv, δw is perpendicular to the plane containing the direction of the vortex-line at (x', y', z') and the line r, and also that its sense is that in which the point (x, y, z) would move if it were rigidly attached to a body rotating with the fluid element at (x', y', z'). The magnitude of the resultant is

$$\{(\delta u)^2 + (\delta v)^2 + (\delta w)^2\}^{\frac{1}{2}} = \frac{\omega \sin \chi}{2\pi r^2} dx'dy'dz' \ldots\ldots\ldots(16),$$

(by an elementary formula of solid geometry), where χ is the angle which r makes with the direction of the vortex-line at (x', y', z').

A relation of exactly the same form as that here developed obtains between the magnetic force and the electric currents in any electro-magnetic field. If we suppose a system of electric currents arranged in exactly the same manner as the vortex-fila-ments, the components of the current at (x', y', z') being ξ, η, ζ, the components of the magnetic force at (x, y, z) due to these currents will be u, v, w.

In the general case (*i.e.* when θ is not everywhere zero) we must add to the values of u, v, w obtained by integrating (15) the terms $\frac{dP}{dx}$, $\frac{dP}{dy}$, $\frac{dP}{dz}$, respectively, where P has the value (10). These are the components of the force at (x, y, z) produced by a distribution of imaginary magnetic matter with density θ.

131.   Let us revise the investigation of Art. 129 with a view to adapting it to the case where the region occupied by the fluid is not infinite, but is limited by surfaces at which the value of the normal velocity λ is given.   The equations to be satisfied by u, v, w are (3), (4) and (5).   The integrals (10) and (11) being supposed to refer to this limited region, the surface-integral in the last member of (12) will not in general vanish unless all the vortices present form closed filaments lying wholly in the region.   If on the other hand the vortex-lines traverse the region, beginning and ending on the boundary, we may suppose them continued outside the region, or along its surface, in such a manner that they form closed curves.   We thus obtain a larger region in which all the vortex-filaments are closed, and if we now suppose the integrals in (10) and (11) to refer to this extended region, the surface-inte-

gral in question will still vanish. On this understanding then the relation (8) is satisfied, and the values of u, v, w thus derived (which we shall now distinguish by a new suffix) will satisfy (3) and (4).

They will not however in general satisfy the boundary-condition (5). Let λ_0 be the value of the normal velocity which the formulæ (6) would give, viz.

$$\lambda_0 = lu_0 + mv_0 + nw_0,$$

and let us write

$$u = u_0 + u_1, \quad v = v_0 + v_1, \quad w = w_0 + w_1,$$

where u_1, v_1, w_1 remain to be found. Substituting in (3), (4) and (5) we obtain

$$\frac{du_1}{dx} + \frac{dv_1}{dy} + \frac{dw_1}{dz} = 0,$$

$$\frac{dw_1}{dy} - \frac{dv_1}{dz} = 0, \quad \frac{du_1}{dz} - \frac{dw_1}{dx} = 0, \quad \frac{dv_1}{dx} - \frac{du_1}{dy} = 0;$$

with the boundary condition

$$lu_1 + mv_1 + nw_1 = \lambda - \lambda_0.$$

Hence we may write

$$u_1 = \frac{dQ}{dx}, \qquad v_1 = \frac{dQ}{dy}, \qquad w_1 = \frac{dQ}{dz},$$

where Q is a single-valued function satisfying

$$\nabla^2 Q = 0 \dots\dots\dots\dots\dots\dots\dots\dots\dots(17)$$

throughout the (simply-connected) region, and making

$$\frac{dQ}{dn} = \lambda - \lambda_0 \dots\dots\dots\dots\dots\dots\dots(18)$$

at the boundary. The problem of finding Q so as to satisfy these conditions was shewn in Art. 49 to be determinate.

Vortex-sheets.

132. We have so far assumed u, v, w to be continuous. We will now shew how cases where these functions are discontinuous may be brought within the scope of our theorems.

Let us suppose that we have a series of vortex-filaments arranged in a thin film over a surface S, and let ω be the angular velocity, and ϵ the thickness, at any point of such a film. Let us examine the form which our previous results assume when ω is increased, and ϵ diminished, without limit, yet in such a way that the product $\omega\epsilon$, ($=\omega'$, say,) remains finite. The infinitely thin film is then called a 'vortex-sheet.'

The functions L, M, N will now consist in part of potentials of matter distributed with surface densities $\dfrac{\xi\epsilon}{2\pi}$, $\dfrac{\eta\epsilon}{2\pi}$, $\dfrac{\zeta\epsilon}{2\pi}$ over S. We know from the theory of Attractions that L, M, N are continuous even when the point to which they refer crosses S, but that their derivatives are discontinuous; viz. the derivative taken in the direction of the normal (drawn in the direction of crossing) experiences an abrupt decrease of amount $4\pi \times$ surface-density. Hence the changes (diminutions) in the values of $\dfrac{dL}{dx}$, $\dfrac{dL}{dy}$, $\dfrac{dL}{dz}$ will be $2l\xi\epsilon$, $2m\xi\epsilon$, $2n\xi\epsilon$, if l, m, n be the direction-cosines of the normal drawn as just explained. The values of u, v, w obtained from (6) will therefore be discontinuous at S, the components of the relative velocity of the portions of fluid on opposite sides of S being

$$2\,(m\zeta - n\eta)\,\epsilon, \qquad 2(n\xi - l\zeta)\,\epsilon, \qquad 2\,(l\eta - m\xi)\,\epsilon\ldots\ldots(19),$$

respectively. These are the amounts by which the components on the side towards which the normal (l, m, n) is drawn fall short of those on the other. This relative velocity is tangential to S, and perpendicular to the vortex-lines. Its amount is $2\omega\epsilon$, or $2\omega'$, and its direction is that due to a rotation of the same sign as ω' about the vortex-lines in the adjacent part of S.

Hence a surface of discontinuity at which the relation

$$lu_1 + mv_1 + nw_1 = lu_2 + mv_2 + nw_2\ldots\ldots\ldots\ldots(20),$$

[(13) of Art. 10] is satisfied may be treated as a vortex-sheet, in which the vortex-lines are everywhere perpendicular to the direction of relative motion of the fluid on the two sides of the surface, and the product ω' of the (infinite) angular velocity into the (infinitely small) thickness is equal to half the amount of this relative velocity.

In the same way, a discontinuity of *normal* velocity is obtained by supposing θ to be infinite throughout a thin film, but in such a way that the product (θ' say) of θ into the thickness ϵ is finite. The normal velocities at adjacent points on opposite sides of the film will then differ by θ'.

Velocity-Potential due to a Vortex.

133. At points external to the vortices there exists of course a velocity-potential, whose value may be found by integration of (15), as follows. Taking, for shortness, the case of a single closed vortex, we write $dx'dy'dz' = \sigma'ds'$, where ds' is an element of the length of the filament, σ' its section. Also we may write

$$\xi = \frac{\omega'dx'}{ds'}, \qquad \eta' = \frac{\omega'dy'}{ds'}, \qquad \zeta = \frac{\omega'dz'}{ds'},$$

so that
$$u = \frac{\omega'\sigma'}{2\pi}\int\left(\frac{d}{dz'}\frac{1}{r}\,dy' - \frac{d}{dy'}\frac{1}{r}\,dz'\right)\dots\dots\dots\dots(21),$$

where the product $\omega'\sigma'$, the strength of the vortex, being constant, is placed outside the sign of integration, which is taken right round the filament. Now the analytical theorem (7) of Art. 40 enables us to replace a line-integral taken round a closed curve by a surface-integral taken over any surface bounded by that curve. To apply this to our case, we write, in the formula cited,

$$u = 0, \qquad v = \frac{d}{dz'}\frac{1}{r}, \qquad w = -\frac{d}{dy'}\frac{1}{r},$$

which give

$$\frac{dw}{dy'} - \frac{dv}{dz'} = -\left(\frac{d^2}{dy'^2} + \frac{d^2}{dz'^2}\right)\frac{1}{r} = \frac{d^2}{dx'^2}\frac{1}{r},$$

$$\frac{du}{dz'} - \frac{dw}{dx'} = \frac{d^2}{dx'dy'}\frac{1}{r},$$

$$\frac{dv}{dx'} - \frac{du}{dy'} = \frac{d^2}{dx'dz'}\frac{1}{r}.$$

Hence (21) becomes
$$u = \frac{\omega'\sigma'}{2\pi}\iint\left(l\frac{d}{dx'} + m\frac{d}{dy'} + n\frac{d}{dz'}\right)\frac{d}{dx'}\frac{1}{r}\,dS',$$

or, since
$$\frac{d}{dx'}\frac{1}{r} = -\frac{d}{dx}\frac{1}{r},$$

$$u = \frac{d\phi}{dx},$$

where

$$\phi = -\frac{\omega'\sigma'}{2\pi} \iint \left(l \frac{d}{dx'} + m \frac{d}{dy'} + n \frac{d}{dz'} \right) \frac{1}{r} \, dS' \dots\dots\dots(22).$$

Here l, m, n denote the normal to the element dS' of a surface bounded by the vortex-filament.

The equation (22) may be otherwise written

$$\phi = -\frac{\omega'\sigma'}{2\pi} \iint \frac{\cos\vartheta}{r^2} \, dS' \dots\dots\dots\dots\dots(23),$$

where ϑ denotes the angle between r and the normal l, m, n. Since $\dfrac{\cos\vartheta . dS'}{r^2}$ is the elementary solid angle subtended by dS' at (x, y, z), we see that the velocity-potential at any point due to a single re-entrant vortex is equal to the product of $-\dfrac{\omega'\sigma'}{2\pi}$ into the solid angle which any surface bounded by the vortex subtends at that point.

Since this solid angle changes by 4π when the point in question describes a circuit embracing the vortex, the value of ϕ given by (23) is cyclic, the cyclic constant being twice the strength of the vortex.   Compare Art. 127.

Dynamical Theorems.

134.   In the theorems which follow, we assume that the external impressed forces have a single-valued potential V, and that ρ is either a constant or a function of p only.

We first consider any terminated line AB drawn in the fluid, and suppose every point of this line to move with the velocity of the fluid at that point. In other words the line moves so as to consist always of the same chain of particles. We proceed to calculate the rate at which the flow along this line, from A to B, is increasing. If dx, dy, dz be the projections on the axes of co-ordinates of an element of the line, we have, with our previous notation,

$$\frac{\partial}{\partial t} (u\,dx) = \frac{\partial u}{\partial t} \, dx + u \frac{\partial dx}{\partial t} .$$

Now $\dfrac{\partial dx}{\partial t}$, the rate at which dx is increasing in consequence of the motion of the fluid, is evidently equal to the difference of the velocities parallel to x at its two ends, i.e. to du; and the value of $\dfrac{\partial u}{\partial t}$ is given in Art. 6. Hence, and by similar considerations, we find

$$\frac{\partial}{\partial t}(udx + vdy + wdz) = -\frac{dp}{\rho} - dV + udu + vdv + wdw.$$

Integrating along the line, from A to B, we get

$$\frac{\partial}{\partial t}\int_A^B (udx + vdy + wdz) = \left[-\int\frac{dp}{\rho} - V + \tfrac{1}{2}q^2\right]_A^B \cdots(24),$$

or, the rate at which the flow from A to B is increasing is equal to the excess of the value which $\tfrac{1}{2}q^2 - V - \int\dfrac{dp}{\rho}$ has at B over that which it has at A.

This theorem, which is due to Thomson, comprehends the whole of the dynamics of a perfect fluid in the general case, as equation (3) of Art. 25 does for the particular case of irrotational motion. For instance, equations (26) of Chapter I. may be derived from it by taking as the line AB the infinitely short line whose projections were originally da, db, dc, and equating separately to zero the coefficients of these infinitesimals.

The expression within brackets on the right-hand side of (24) is a single-valued function of x, y, z. It follows that if the integration on the left-hand side be taken round a closed curve, (so that B coincides with A,) we have

$$\frac{\partial}{\partial t}\int (udx + vdy + wdz) = 0,$$

or, the circulation in any circuit moving with the fluid does not alter with the time. See Art. 59.

Applying this theorem to a circuit embracing a vortex-tube we find that the strength of any vortex is constant.

Also, remembering the formula given in Art. 39 for the circulation in an infinitesimal circuit, we see that if throughout any portion of a fluid mass in motion the conditions $\xi = 0$, $\eta = 0$, $\zeta = 0$ obtain at any one instant, the same is true for the same portion of the mass at every other instant, which is the theorem of Art. 23.

It follows that rotational motion cannot be produced in any part of a fluid mass by the action of forces which have a single-valued potential, and that such a motion, if already existent, cannot be destroyed by the action of such forces.

If we take at any instant a surface composed wholly of vortex-lines, the circulation in any circuit drawn on it is zero, by Art. 40, for we have $l\xi + m\eta + n\zeta = 0$ at every point of the surface. The preceding article shews that if the surface be now supposed to move with the fluid, the circulation will always be zero in any circuit drawn on it, and therefore the surface will always consist of vortex-lines. Again, considering two such surfaces, it is plain that their intersection must always be a vortex-line, whence we derive the theorem that the vortex-lines move with the fluid.

This remarkable theorem was first given by Helmholtz* for the case of liquids; the preceding proof, by Thomson, shews it to be applicable to all fluids satisfying the conditions stated at the beginning of this article.

Kinetic Energy.

135. The formula for the kinetic energy, viz.

$$2T = \iiint \rho(u^2 + v^2 + w^2)\,dx\,dy\,dz \dots\dots\dots(25),$$

may be put into several remarkable and useful forms. We confine ourselves, for simplicity, to the case where the fluid (supposed incompressible) extends to infinity and is at rest there, and where further all the vortices present are within a finite distance of the origin.

We have in this case, $\theta = 0$, $P = 0$, $\rho = $ const., so that (25) becomes on substitution from (6),

$$2T = \rho \iiint \left\{ u\left(\frac{dN}{dy} - \frac{dM}{dz}\right) + v\left(\frac{dL}{dz} - \frac{dN}{dx}\right) + w\left(\frac{dM}{dx} - \frac{dL}{dy}\right) \right\} dx\,dy\,dz.$$

* See note (D).

This triple integral may, exactly as in Art. 129 (12), be replaced by the sum of a surface-integral

$$\rho \iint \{L\,(nv - mw) + M\,(lw - nu) + N\,(mu - lv)\}\,dS \quad \ldots\ldots(26),$$

and a volume-integral

$$\rho \iiint \left\{ L \left(\frac{dw}{dy} - \frac{dv}{dz} \right) + M \left(\frac{du}{dz} - \frac{dw}{dx} \right) + N \left(\frac{dv}{dx} - \frac{du}{dy} \right) \right\} dx\,dy\,dz$$
$$\ldots\ldots\ldots(27).$$

Now it appears from (11) that at an infinite distance R from the origin, L, M, N are at most* of the order $\dfrac{1}{R}$, and therefore u, v, w at most of the order $\dfrac{1}{R^2}$, whereas when the external bounding surface is increased in all its dimensions without limit the surface-elements dS increase proportionately to R^2 only. The surface-integral (26) is therefore of an order not higher than $\dfrac{1}{R}$, and therefore vanishes in the limit. Hence

$$T = \rho \iiint (L\xi + M\eta + N\zeta)\,dx\,dy\,dz \ldots\ldots\ldots\ldots(28).$$

If we substitute the values of L, M, N from (11), this becomes

$$T = \frac{\rho}{2\pi} \iiint\!\!\iiint \frac{\xi\xi' + \eta\eta' + \zeta\zeta'}{r}\,dx\,dy\,dz\,dx'dy'dz' \ldots\ldots(29),$$

where each of the volume integrations extends over all the vortices.

136. Under the same circumstances we have another useful expression for T; viz.

$$T = 2\rho \iiint \{u\,(y\zeta - z\eta) + v\,(z\xi - x\zeta) + w\,(x\eta - y\xi)\}\,dx\,dy\,dz \ldots(30).$$

To verify this, we take the right-hand member, and transform it by the process already so often employed, omitting the surface-integrals for the same reason as in the preceding article. The first of the three terms gives

---

* They are in fact of the order $\dfrac{1}{R^2}$, as may be seen (for example) by calculating the value of L for a single closed vortex, and expressing it, by the method of Art. 133, as a surface-integral taken over a surface bounded by the vortex. Consequently the velocities u, v, w are really of the order $\dfrac{1}{R^3}$.

$$\rho \iiint u \left\{ y \left(\frac{dv}{dx} - \frac{du}{dy} \right) - z \left(\frac{du}{dz} - \frac{dw}{dx} \right) \right\} dx\,dy\,dz$$

$$= -\rho \iiint \left\{ (vy + wz) \frac{du}{dx} - u^2 \right\} dx\,dy\,dz.$$

Transforming the remaining terms in the same way, adding, and making use of the equation of continuity

$$\frac{du}{dx} + \frac{dv}{dy} + \frac{dw}{dz} = 0,$$

we obtain

$$\rho \iiint \left(u^2 + v^2 + w^2 + xu \frac{du}{dx} + yv \frac{dv}{dy} + zw \frac{dw}{dz} \right) dx\,dy\,dz,$$

or, finally, on again transforming the last three terms,

$$\tfrac{1}{2} \rho \iiint (u^2 + v^2 + w^2) dx\,dy\,dz,$$

i.e. T.

The value (30) of T must of course be unaltered by any displacement of the axes of co-ordinates. This consideration gives

$$\left. \begin{aligned} \iiint (w\eta - v\zeta)\,dx\,dy\,dz &= 0 \\ \iiint (u\zeta - w\xi)\,dx\,dy\,dz &= 0 \\ \iiint (v\xi - u\eta)\,dx\,dy\,dz &= 0 \end{aligned} \right\} \dots\dots\dots\dots\dots (31),$$

relations which of course admit of independent verification. Thus

$$2 \iiint (w\eta - v\zeta)\,dx\,dy\,dz = \iiint \left\{ w \left(\frac{du}{dz} - \frac{dw}{dx} \right) - v \left(\frac{dv}{dx} - \frac{du}{dy} \right) \right\} dx\,dy\,dz$$

$$= -\iiint \left(u \frac{dw}{dz} + u \frac{dv}{dy} \right) dx\,dy\,dz = \iiint u \frac{du}{dx}\,dx\,dy\,dz = 0,$$

and similarly for the others.

137. The rate at which the energy of any mass of liquid is increasing at any instant is

$$\frac{dT}{dt} = \rho \iiint \left(u \frac{\partial u}{\partial t} + v \frac{\partial v}{\partial t} + w \frac{\partial w}{\partial t} \right) dx\,dy\,dz$$

$$= -\iiint \left(u \frac{dp}{dx} + \dots + \rho u \frac{dV}{dx} + \dots \right) dx\,dy\,dz$$

$$= \iint (p + \rho V)(lu + mv + nw)\,dS \dots\dots\dots\dots\dots (32),$$

if l, m, n be the direction cosines of the inwardly-directed normal to any element dS of the boundary. The part of this expression which contains p gives the rate at which the external pressure works, the remaining part expresses the rate at which the mass is losing potential energy. If the mass be enclosed within fixed rigid walls, we have

$$lu + mv + nw = 0$$

at the boundary, and therefore $\dfrac{dT}{dt} = 0$, or $T = \text{const.}$ The same result holds for the case of an unlimited mass of liquid subject to the conditions of Art. 129. We then have, beyond the vortices

$$p + \rho V = -\rho \frac{d\phi}{dt} - \tfrac{1}{2}\rho q^2,$$

and it appears from Chapter III. that at an infinite distance from the origin ϕ, and therefore also $\dfrac{d\phi}{dt}$ is constant with respect to x, y, z. Under these circumstances the surface integral in (32) is zero. Compare Art. 65.

138. We proceed to apply the foregoing general theory to the discussion of some simple cases.

1. *Rectilinear Vortices.*

Suppose that we have an infinite mass of liquid in motion in two dimensions (xy), so that u, v are functions of x, y only, and $w = 0$. We have then $\xi = 0$, $\eta = 0$ everywhere and therefore also $L = 0$, $M = 0$. The value of N is

$$N = \frac{1}{2\pi} \iiint \frac{\zeta' dx' dy' dz'}{r},$$

and if we perform the integration with respect to z' between the limits $\pm \gamma$, and then make γ infinite, we find

$$N = \frac{1}{2\pi} \log 4\gamma^2 \iint \zeta' dx' dy' - \frac{1}{\pi} \iint \zeta' \log r \, dx' dy',$$

where r now (and as far as Art. 140) stands for $\{(x - x')^2 + (y - y')^2\}^{\frac{1}{2}}$. The first term in the value of N, though infinite, is constant, and since we are concerned only with the differential coefficients of N, we may write

$$N = -\frac{1}{\pi} \iint \zeta' \log r \, dx' dy' \dots \dots \dots \dots \dots (33).$$

The formulæ (6) then give

$$u = \frac{dN}{dy} = -\frac{1}{\pi} \iint \zeta' \frac{y - y'}{r^2} dx' dy',$$

$$v = -\frac{dN}{dx} = \frac{1}{\pi} \iint \zeta' \frac{x - x'}{r^2} dx' dy'.$$

We see that N is identical with the function ψ of Chapter IV.

A vortex-filament whose co-ordinates are x', y' and strength m' contributes to the motion at (x, y) a velocity whose components are

$$-\frac{m'}{\pi} \cdot \frac{y - y'}{r^2}, \text{ and } \frac{m'}{\pi} \cdot \frac{x - x'}{r^2}.$$

This velocity is perpendicular to the line joining the points (x, y) (x', y'), and its amount is $\dfrac{m'}{\pi r}$.

Let us calculate the integrals $\iint u\zeta dx dy$, and $\iint v\zeta dx dy$, where the integrations include all portions of the plane xy for which ζ does not vanish. We have

$$\iint u\zeta dx dy = -\frac{1}{\pi} \iiiint \zeta\zeta' \frac{y - y'}{r^2} dx dy \, dx' dy',$$

where each double integration includes the sections of all the vortices. Now, corresponding to any term

$$\zeta\zeta' \frac{y - y'}{r^2} dx dy \, dx' dy'$$

of this integral, we have another term

$$\zeta\zeta' \frac{y' - y}{r^2} dx dy \, dx' dy',$$

and these terms neutralize one another. Hence

$$\iint u\zeta dx dy = 0 \dots \dots \dots \dots \dots (34);$$

and, by the same reasoning,

$$\iint v\zeta dx dy = 0 \dots \dots \dots \dots \dots (35).$$

If we denote as before the strength of a vortex by m, these results may be written

$$\Sigma mu = 0, \quad \Sigma mv = 0 \dots\dots\dots\dots\dots(36).$$

We have seen above that the strength of each vortex is constant with regard to the time. Hence (36) express that the point whose co-ordinates are

$$\bar{x} = \frac{\Sigma mx}{\Sigma m}, \qquad \bar{y} = \frac{\Sigma my}{\Sigma m},$$

is fixed throughout the motion. This point, which coincides with the centre of inertia of a mass distributed over the plane xy with the surface-density ζ, may be called the 'centre' of the system of vortices, and the straight line parallel to z of which it is the projection may be called the 'axis' of the system.

139.   We proceed to discuss some particular cases.

(a)   First, let us suppose that we have only one vortex-filament present, and let ζ have the same sign throughout its infinitely small section. Its centre, as just defined, will lie either within the substance of the filament, or at all events infinitely close to it. Since this centre remains at rest, the filament as a whole will be stationary, though its parts may experience relative motions, and its centre will not necessarily lie always in the same element of fluid. Any particle at a finite distance r from the centre of the filament will describe a circle about the latter as axis, with constant velocity $\frac{m}{\pi r}$. The region external to the filament is doubly-connected; and the circulation in any (simple) circuit embracing the filament is $2m$. The irrotational motion of the fluid external to the filament is the same as in Art. 35.

(b)   Next suppose that we have two vortices, of strengths m_1, m_2, respectively. Let A, B be their centres, O the centre of the system. The motion of each filament is entirely due to the other filament, and is therefore always perpendicular to AB. Hence the two filaments remain always at the same distance from one another, and rotate with uniform angular velocity about O, which is fixed. This angular velocity is easily

found; we have only to divide the velocity of A (say), viz.
$\dfrac{m_2}{\pi \cdot AB}$, by the distance AO, where

$$AO = \frac{m_2}{m_1 + m_2} AB,$$

and so obtain $\dfrac{m_1 + m_2}{\pi \cdot AB^2}$ for the value required.

If m_1, m_2 be of the same sign, i.e. if the directions of rotation in the two filaments be the same, O lies between A and B; but if the directions be of opposite signs, O lies in AB, or BA, produced.

If $m_1 = -m_2$, O is at infinity; in this case it is easily seen that A, B move with uniform velocity $\dfrac{m_1}{\pi \cdot AB}$ perpendicular to AB, which remains fixed in direction. The motion external to the filaments at any instant is given by the formulæ of Chapter IV, Example 3.

The motion at all points of the plane bisecting AB at right angles is tangential to that plane. We may therefore suppose this plane to form a fixed rigid boundary of the fluid on either side of it; and so obtain the solution of the case where we have a single rectilinear vortex in the neighbourhood of a fixed plane wall to which it is parallel. The filament moves parallel to the plane with the velocity $\dfrac{m}{2\pi d}$, where d is the distance of the vortex from the wall.

In the last case $[m_1 = -m_2]$ the stream-lines are all circles. We can hence derive the solution of the case where we have a single vortex-filament in a mass of fluid which is bounded, either internally or externally, by a fixed circular cylinder. Thus, in

Fig. 11.

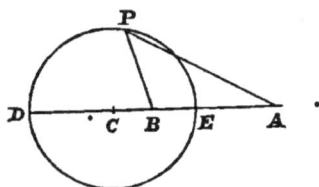

the figure, let DPE be the section of the cylinder, A the position of the vortex (supposed in this case external), and let B be the 'image' of A with respect to the circle DPE, viz. C being the centre, let

$$CB . CA = c^2,$$

where c is the radius of the circle. If P be any point on the circle, we have

$$\frac{AP}{BP} = \frac{AE}{BE} = \frac{AD}{BD} = \text{const.} ;$$

so that the circle occupies the position of a stream-line due to a pair of vortices, whose strengths are equal and opposite in sign, situated at A, B in an unlimited mass of fluid. Since the motion of the vortex A would then be perpendicular to AB, it is plain that all the conditions of the problems are satisfied if we suppose A to describe a circle about the axis of the cylinder with uniform velocity

$$\frac{m}{\pi . AB} = \frac{m . CA}{\pi (CA^2 - c^2)} .$$

In the same way a single vortex of strength m, situated within a fixed circular cylinder, say at B, would describe a circle with uniform velocity $\dfrac{m . CB}{\pi (c^2 - CB^2)}$.

(c) If we have four parallel rectilinear vortices whose centres form a rectangle $ABB'A'$, the strengths being m for the vortices A', B, and $-m$ for the vortices A, B', it is evident that the centres will always form a rectangle. Further, if the various rotations have the directions indicated in the figure, we see that

Fig. 12.

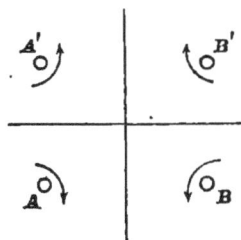

the effect of the presence of the pair A', B' on A, B is to separate them, and at the same time to diminish their velocity perpendicular to the line joining them. The planes which bisect AB,

AA' at right angles may (either or both) be taken as fixed rigid boundaries. We thus get the case where a pair of vortices, of equal and opposite strengths, move towards (or from) a plane wall, or where a single vortex moves in the angle between two perpendicular walls.

For other interesting cases of motion of rectilinear vortices we refer to a paper by Professor Greenhill*.

2. Circular Vortices.

140. Next let us take the case where all the vortices present in the fluid (supposed unlimited as before) are circular, having the axis of x as a common axis. Let ϖ denote the distance of any point P from this axis, ϑ the angle which ϖ makes with the plane xy, v the velocity in the direction of ϖ, and ω the angular velocity of the element at P. It is evident that u, v, ω are functions of x and ϖ only, and that the axis of ω is perpendicular to x, ϖ. We have then

$$\left. \begin{array}{ll} y = \varpi \cos \vartheta, & z = \varpi \sin \vartheta, \\ v = v \cos \vartheta, & w = v \sin \vartheta, \\ \xi = 0, \quad \eta = -\omega \sin \vartheta, & \zeta = \omega \cos \vartheta \end{array} \right\} \dots \dots \dots (37).$$

If we make these substitutions, writing $\varpi d\vartheta\, dx\, d\varpi$ for the volume-element, in (30), and perform the integration with respect to ϑ, we obtain

$$T = 4\pi\rho \iint (\varpi u - xv)\, \varpi\omega\, dx\, d\varpi.$$

The second and third of equations (31) are satisfied identically; the first gives

$$\iint \varpi v \omega\, dx\, d\varpi = 0.$$

If we denote by m the strength $\omega dx d\varpi$ of the vortex whose co-ordinates are x, ϖ, these results may be written

$$\Sigma m (\varpi u - xv)\, \varpi = \frac{T}{4\pi\rho} \dots \dots \dots \dots (38),$$

$$\Sigma m \varpi v = 0 \dots \dots \dots \dots \dots (39),$$

---

* Quarterly Journal of Mathematics, 1877.

where the summations embrace all the vortices present in the fluid. If in these equations we suppose x, ϖ always to refer to the same vortex, we may write

$$u = \frac{dx}{dt}, \quad v = \frac{d\varpi}{dt}.$$

Since m is constant for the same vortex, the equation (39) is at once integrable with respect to t, whence

$$\Sigma m \varpi^2 = \text{const} \dots\dots\dots\dots\dots\dots\dots(40).$$

A quantity ϖ_0 defined by the equation

$$\varpi_0^2 = \frac{\Sigma m \varpi^2}{\Sigma m} \dots\dots\dots\dots\dots\dots\dots(41),$$

may be called the 'mean radius' of the vortex-rings. The equation (40) shews that this mean radius is constant throughout the motion.

If we introduce in addition a magnitude x_0 such that

$$x_0 \Sigma m \varpi^2 = \Sigma m \varpi^2 x \dots\dots\dots\dots\dots(42),$$

it is plain that the position of the circle whose co-ordinates are x_0, ϖ_0 depends only on the strengths and the configuration of the vortices, and not on the position of the origin of co-ordinates. This circle may be called the 'circular axis' of the whole system of vortex-rings. It remains constant in radius; and its motion parallel to x is obtained by differentiating (42), viz. we have

$$\frac{dx_0}{dt} \Sigma m \varpi^2 = \Sigma m \varpi^2 \frac{dx}{dt} + 2\Sigma m x \varpi \frac{d\varpi}{dt},$$

or, by (38) and (41);

$$\Sigma m \varpi_0^2 \frac{dx_0}{dt} = \frac{T}{4\pi\rho} + 3\Sigma m (x - x_0) \varpi \frac{d\varpi}{dt} \dots\dots(43),$$

where we have added to the right-hand side a term which vanishes in virtue of (40).

141. The formulæ (11) become, on making the substitutions (37),

$$L = 0$$

$$M = -\frac{1}{2\pi} \iiint \frac{\omega' \sin \vartheta'}{r} \varpi' d\vartheta' dx' d\varpi',$$

$$N = +\frac{1}{2\pi} \iiint \frac{\omega' \cos \vartheta'}{r} \varpi' d\vartheta' dx' d\varpi',$$

where

$$r = \{(x - x')^2 + \varpi^2 + \varpi'^2 - 2\varpi\varpi' \cos(\vartheta - \vartheta')\}^{\frac{1}{2}}.$$

Now if

$$S = \frac{1}{2\pi} \iiint \frac{\varpi' \cos(\vartheta - \vartheta')}{r} \varpi' d\vartheta' dx' d\varpi' \ldots\ldots\ldots(44),$$

we have

$$S = -M \sin\vartheta + N \cos\vartheta \ldots\ldots\ldots\ldots(45);$$

and, since the integral

$$\iiint \frac{\varpi' \sin(\vartheta - \vartheta')}{r} \varpi' d\vartheta' dx' d\varpi'$$

is identically zero,

$$0 = M \cos\vartheta + N \sin\vartheta \ldots\ldots\ldots\ldots\ldots(46).$$

Combining (45) and (46), we find

$$M = -S \sin\vartheta, \quad N = S \cos\vartheta \ldots\ldots\ldots\ldots(47).$$

If the variable of integration in (44) be changed from ϑ' to ϵ, where $\epsilon = \vartheta' - \vartheta$, the limits of integration for ϵ are 0 and 2π; and since

$$\frac{\cos\epsilon}{\{(x - x')^2 + \varpi^2 + \varpi'^2 - 2\varpi\varpi' \cos\epsilon\}^{\frac{1}{2}}}$$

$$= \left\{ \frac{(x - x')^2 + (\varpi + \varpi')^2}{2\varpi\varpi'} - 1 \right\} \frac{1}{r} - \frac{r}{2\varpi\varpi'};$$

and

$$r^2 = \{(x - x')^2 + (\varpi + \varpi')^2\} \left\{ 1 - \frac{4\varpi\varpi'}{(x - x')^2 + (\varpi + \varpi')^2} \cos^2\frac{\epsilon}{2} \right\},$$

we may write (44) in the form

$$S = \frac{1}{\pi} \iint \sqrt{\frac{\varpi'}{\varpi}} \left\{ \left(\frac{2}{k} - k \right) F_1 - \frac{E_1}{k} \right\} \varpi' dx' d\varpi' \ldots\ldots(48).$$

Here F_1, E_1 denote the complete elliptic integrals of the first and second kinds with respect to the modulus

$$k = \left\{ \frac{4\varpi\varpi'}{(x - x')^2 + (\varpi + \varpi')^2} \right\}^{\frac{1}{2}} \ldots\ldots\ldots\ldots(49),$$

viz.,

$$F_1 = \int_0^{\frac{\pi}{2}} \frac{d\theta}{\sqrt{1 - k^2 \sin^2\theta}}, \quad E_1 = \int_0^{\frac{\pi}{2}} \sqrt{1 - k^2 \sin^2\theta}\, d\theta.$$

142. The kinds of fluid motion now under consideration come under the class for which a stream-function ψ was shewn, in Art. 103, to exist. By the definition of that article, we have

$2\pi\psi =$ total flux through the circle whose co-ordinates are x, ϖ,

$\quad = \iint(lu + mv + nw)\,dS$,

where the integration extends over any surface bounded by that circle. Recalling the expressions (6) for u, v, w, from which P is now to be omitted, we have by the theorem of Art. 40,

$$2\pi\psi = \int(L\,dx + M\,dy + N\,dz),$$

the integration here being taken round the circle, or, by (47),

$$\psi = \varpi S \quad\dots\dots\dots\dots\dots\dots\dots(50).$$

The formula (28) for the kinetic energy may now be written

$$T = 2\pi\rho \iint S'\omega'\varpi'\,dx'\,d\varpi'$$

$$\quad = 2\pi\rho \iint \psi'\omega'\,dx'\,d\varpi'\dots\dots\dots\dots\dots(51).$$

143. Let us take the case of a single circular vortex of strength m. At all points of its infinitely small section the modulus k of the elliptic integrals in the value of S is nearly equal to unity. In this case we have*

$$F_1 = \log\frac{4}{k'}, \qquad E_1 = \tfrac{1}{2}\pi,$$

approximately, where k' denotes the complementary modulus $\sqrt{(1 - k^2)}$, so that in our case

$$k'^2 = \frac{(x - x')^2 + (\varpi - \varpi')^2}{(x - x')^2 + (\varpi + \varpi')^2} = \frac{\delta^2}{4\varpi^2},$$

nearly, if δ denote the distance between the infinitely near points (x, ϖ), (x', ϖ'). Hence at points within the substance of the vortex the value of S, and therefore by (50) also of ψ, is of the order $m \log \epsilon$, where ϵ is a small linear magnitude comparable with the dimensions of the section. The velocity at the same point, depending (Art. 103) on the differential coefficient of ψ, will be of the order $\dfrac{m}{\epsilon}$.

* See Cayley, *Elliptic Functions*, Art. 72, and Maxwell, *Electricity and Magnetism*, Arts. 704, 705.

We can now make use of (43) to estimate the magnitude of the velocity $\dfrac{dx_0}{dt}$ of translation of the vortex. By (51) T is of the order $m^2 \log \epsilon$, and $\dfrac{d\varpi}{dt}$ is, as we have seen, of the order $\dfrac{m}{\epsilon}$. Also $x - x_0$ is of the order ϵ. Hence the second term on the right-hand side of (43) is, in this case, small compared with the first, and the velocity of translation of the ring is of the order $m \log \epsilon$, and approximately constant.

An isolated vortex-ring moves then, without sensible change of size, parallel to its (rectilinear) axis, with nearly constant velocity. This velocity is small compared with that of the fluid in the immediate neighbourhood of its circular axis, but large compared with $\dfrac{2\pi m}{\varpi'}$, the velocity of the fluid at the centre of the ring, with which it agrees in direction.

A drawing of the stream-lines due to a single circular vortex is given by Thomson[*].

144. If we have any number of circular vortex-rings, coaxial or not, the motion of any one of these may be conceived as made up of two parts, one due to the ring itself, the other due to the influence of the remaining rings. The preceding considerations shew that the second part is insignificant compared with the first, except when two or more rings approach within a very small distance of one another. Hence each ring will move, without sensible change of shape or size, with nearly uniform velocity in the direction of its (rectilinear) axis, until it passes within a short distance of a second ring. A general notion of the result of the encounter of two rings may, in particular cases, be gathered from the theorem of Art. 130.

Thus, let us suppose that we have two circular vortices having the same rectilinear axis. If the sense of the rotation be the same in both, the two rings will advance in the same direction. One effect of their mutual influence will be to increase the radius

---

[*] On Vortex-Motion, *Trans. R. S. Edin.* 1869. Copied in Maxwell, *Electricity and Magnetism*, plate xviii.

of the one in advance, and to contract the radius of the one in
the rear. If the radius of the one in front become larger than
that of the one in the rear, the motion of the former ring will
be retarded, whilst that of the latter is accelerated. Hence if
the conditions as to the relative size and strength of the two
rings be favourable, it may happen that the second ring will
overtake and pass through the first. The parts played by the
two rings will then be reversed; the one which is now in the
rear will in turn overtake and pass through the other, and so on,
the rings alternately passing one through the other.

If the rotations in the two rings be opposed, and such that
the rings approach one another, the mutual influence will be to
enlarge the radius of each ring.

If the two rings be moreover equal in size and strength, the
velocity of approach will continually diminish. In this case the
motion at all points of the plane which is parallel to the two
rings, and half-way between them, is tangential to this plane.
We may therefore, if we please, regard this plane as a fixed
boundary to the fluid on either side of it, and so obtain the
solution of the case where a single vortex-ring moves directly
towards a fixed rigid wall.

On the Conditions for Steady Motion.

145. In steady motion, *i.e.* when

$$\frac{du}{dt} = 0, \quad \frac{dv}{dt} = 0, \quad \frac{dw}{dt} = 0,$$

the equations (2) of Art. 6 may be written

$$u\frac{du}{dx} + v\frac{dv}{dx} + w\frac{dw}{dx} - 2(v\zeta - w\eta) = -\frac{dV}{dx} - \frac{1}{\rho}\frac{dp}{dx}, \text{ &c., &c.}$$

Hence if we make

$$P = \int\frac{dp}{\rho} + V + \tfrac{1}{2}q^2 \dots\dots\dots\dots(52),$$

we have

$$\frac{dP}{dx} = 2(v\zeta - w\eta), \quad \frac{dP}{dy} = 2(w\xi - u\zeta), \quad \frac{dP}{dz} = 2(u\eta - v\xi).$$

It follows that

$$u \frac{dP}{dx} + v \frac{dP}{dy} + w \frac{dP}{dz} = 0,$$

$$\xi \frac{dP}{dx} + \eta \frac{dP}{dy} + \zeta \frac{dP}{dz} = 0,$$

so that each of the surfaces $P = $ const. contains both stream-lines and vortex-lines. If further dn denote an element of the normal at any point of such a surface, we have

$$\frac{dP}{dn} = q\omega \sin \theta \quad \dots\dots\dots\dots\dots\dots\dots(53),$$

where q is the current-velocity, ω the angular velocity, and θ the angle between the stream-line and the vortex-line at that point.

Hence the conditions that a given state of motion of a fluid may be a possible state of steady motion are as follows. It must be possible to draw in the fluid an infinite system of surfaces each of which is covered by a network of stream-lines and vortex-lines; and the product of $q\omega \sin \theta \, dn$ must be constant over each such surface, dn denoting the length of the normal drawn to a consecutive surface of the system.

These conditions may also be deduced from the considerations that the stream-lines are, in steady motion, the actual paths of the particles, that the product of the angular velocity into the cross-section is the same at all points of a vortex, and that this product is, for the same vortex, constant with regard to the time.

The theorem that the quantity P, defined by (52), is constant over each surface of the above kind is an extension of that of Art. 28, where it was shewn that P is constant along a stream-line.

The above conditions are satisfied identically in all cases of irrotational motion.

In the motion of a liquid in two dimensions, the product $q \, dn$ is constant along a stream-line; the conditions then reduce to this, that ω (or ζ, if the axes of co-ordinates be the same as in Chapter IV.) must be constant along a stream-line, *i.e.* by Art. 69 (5),

$$\frac{d^2\psi}{dx^2} + \frac{d^2\psi}{dy^2} = f(\psi) \quad \dots\dots\dots\dots(54),$$

where $f(\psi)$ is an arbitrary function of ψ.

A particular solution of (54) is

$$\psi = \tfrac{1}{2}(Ax^2 + 2Bxy + Cy^2),$$

in which case the stream-lines are a system of similar and coaxial conic sections. The velocities at corresponding points are proportional to the linear dimensions of the curves; the angular velocity at any point is $A + C$, and is therefore uniform.

In the case of motion symmetrical about an axis (say that of x), considered in Art. 103, we have $2\pi\varpi dn\,q$ constant along a stream-line. The conditions for steady motion reduce then to this, that the ratio $\omega : \varpi$ must be constant along a stream-line. If ψ be the stream-function, we have

$$\omega = -\frac{1}{\varpi}\left(\frac{d^2\psi}{dx^2} + \frac{d^2\psi}{d\varpi^2} - \frac{1}{\varpi}\frac{d\psi}{d\varpi}\right) \quad \dots\dots\dots(55),$$

so that the condition is

$$\frac{d^2\psi}{dx^2} + \frac{d^2\psi}{d\varpi^2} - \frac{1}{\varpi}\frac{d\psi}{d\varpi} = \varpi^2 f(\psi) \quad \dots\dots\dots(56)^*,$$

where $f(\psi)$ is an arbitrary function of ψ.

---

* Conditions (54), (56) were given by Stokes, *Camb. Trans.* Vol. VII. (1842).

CHAPTER VII.

146. Any disturbance which is propagated from one part of a medium to another, whilst the particles of the medium do not themselves share in the progressive motion of the disturbance, but only deviate slightly from their original positions, is called a 'wave.'

In the case of a liquid under the action of gravity, the disturbance manifests itself by the production of elevations and depressions which travel over the surface (originally, of course, plane and horizontal). That the particles of the fluid do not follow the motion of the waves, but always remain in the neighbourhood of their undisturbed positions, may in this case be readily ascertained by watching the motion of a small floating body.

The present chapter is devoted to the study of waves in liquids. We shall suppose for the most part that the liquid is uniform in depth, that it is unlimited horizontally, and that the motion takes place in two dimensions (one horizontal, the other vertical), so that the elevations and depressions of the surface present the appearance of a series of parallel straight ridges and furrows.

Waves of Small Vertical Displacement*.

147. We take first the case where the maximum horizontal motion is so large compared with the maximum vertical motion,

---

* Stokes, *Camb. and Dublin Math. Journal*, Vol. IV.

that the vertical motion may be altogether neglected. We shall learn from our results in what cases this assumption is legitimate. Let the origin be taken in the bottom of the fluid, the axis of x horizontal, that of y vertical and upwards; and let us suppose that the motion takes place in these two dimensions x, y. Let h be the depth of the fluid in the undisturbed state, $h + \eta$ the ordinate of the surface corresponding to the abscissa x, at the time t. Since the vertical motion is neglected, the pressure at any point (x, y) will be simply that due to the depth below the surface, viz.

$$p = g\rho(h + \eta - y) + \text{const.}$$

Hence

$$\frac{dp}{dx} = g\rho \frac{d\eta}{dx} \dots\dots\dots\dots\dots\dots\dots\dots\dots(1),$$

which is independent of y, so that the horizontal accelerating force is the same for all particles in a plane perpendicular to x. It follows that all particles which once lie in such a plane always do so. It is convenient, now, to follow the Lagrangian method (Art. 16), changing however the notation. Let $x + \xi$ denote the abscissa at time t of the particles whose undisturbed abscissa is x; we have seen that ξ is in fact a function of x only. Further let the independent variable x in p and η refer always to the same portion of fluid; in which case (1) still holds. The volume of fluid, corresponding to unit breadth originally contained between the two planes x and $x + dx$ is $h\,dx$; at the time $t + dt$ the same stratum of fluid has a thickness $dx + \frac{d\xi}{dx}dx$, and its height is $h + \eta$. The equation of continuity therefore is

$$\left(1 + \frac{d\xi}{dx}\right)(h + \eta) = h \dots\dots\dots\dots\dots(2).$$

The equation of motion of the stratum is

$$\rho h\,dx\frac{d^2\xi}{dt^2} = -\frac{dp}{dx}dx\,(h + \eta).$$

With the help of (1) and (2), this becomes

$$\frac{d^2\xi}{dt^2} = gh\frac{\dfrac{d^2\xi}{dx^2}}{\left(1 + \dfrac{d\xi}{dx}\right)^3} \dots\dots\dots\dots\dots\dots\dots\dots(3).$$

So far the only assumption made is that the vertical motion is small compared with the horizontal. If we now assume in addition that $\frac{d\xi}{dx}$ is also a small quantity, in other words that the relative displacement of the two neighbouring elements of fluid never amounts to more than a small fraction of the distance between them, (2) may be written

$$\frac{\eta}{h} = -\frac{d\xi}{dx} \quad\text{.........................(4)},$$

and (3) becomes

$$\frac{d^2\xi}{dt^2} = gh\,\frac{d^2\xi}{dx^2} \quad\text{.......................(5)}.$$

The elevation η then satisfies an equation of exactly the same form.

The above investigations apply to a straight canal of rectangular section. We can however easily extend them to the case of a canal whose section is any whatever, provided it be uniform. Thus, if A be the area of the section, b the breadth of the canal at the undisturbed level, we have instead of (2)

$$\left(1 + \frac{d\xi}{dx}\right)(A + b\eta) = A,$$

whence

$$\eta = -\frac{A}{b}\,\frac{d\xi}{dx},$$

and

$$\frac{d^2\xi}{dt^2} = \frac{gA}{b}\,\frac{d^2\xi}{dx^2} \quad\text{...........................(6)}.$$

148. If for shortness we write

$$c^2 = gh\,* \quad\text{................................(7)},$$

$$x - ct = x_1, \qquad x + ct = x_2,$$

and transform (5) by making x_1, x_2 the independent variables instead of x, t, the equation becomes

$$\frac{d^2\xi}{dx_1\,dx_2} = 0,$$

* Or $c^2 = \frac{gA}{b}$, in (6).

so that the complete solution of (5) is

$$\xi = F(x - ct) + f(x + ct)\dots\dots\dots\dots\dots(8),$$

where F, f denote arbitrary functions.

The interpretation of (8) is simple. Take first the motion represented by the first term alone. Since $F(x - ct)$ is unaltered when t is increased by τ, and x by $c\tau$, it is plain that the disturbance of the particle x at the time t has been communicated, at the time $t + \tau$ to the particle $x + c\tau$, so that the disturbance advances as a whole with uniform velocity c relative to the fluid. The first term of (8) denotes then a wave travelling in the positive direction, without change of form, with velocity c. The second term denotes in like manner a wave travelling with the same velocity in the direction of x negative. Any motion whatever of the fluid, subject to the restrictions of the preceding article, may be regarded as made up of waves of these two kinds.

The velocity of propagation c is, by (7), that 'due to' half the depth of the undisturbed fluid.

149. Let us examine the motion of a surface-particle as a wave passes over it. To fix the ideas we shall suppose the wave to be one of elevation, so that η is everywhere positive, and to be travelling in the positive direction, so that

$$\xi = F(x - ct)\dots\dots\dots\dots\dots\dots\dots(9).$$

We shall also suppose the length λ of the wave to be finite. By differentiation of (9) we find

$$\frac{d\xi}{dt} = -c\frac{d\xi}{dx},$$

or, by (4),

$$\frac{d\xi}{dt} = \frac{c}{h}\eta \dots\dots\dots\dots\dots\dots\dots(10).$$

The particle remains at rest until it is reached by the wave; it then moves forwards with a horizontal velocity proportional to its elevation above the mean level. Also

$$\xi = \frac{1}{h}\int\eta\, cdt,$$

so that the total horizontal displacement at any time is equal to the whole volume (per unit breadth) of elevated fluid which has up to that time passed over the particle, divided by the depth of the fluid. The horizontal motion is therefore at first infinitely small compared with the vertical motion, so that the particle moves at first upwards; its horizontal velocity gradually increases, and attains a maximum when the highest part of the wave is passing it, the vertical motion being then zero. The particle then begins to fall, its horizontal motion at the same time slackening, until it finally comes to rest at its original level, but in advance of its former position by a distance

$$= \frac{\text{total volume of elevation, per unit breadth}}{\text{depth of fluid}}.$$

If the wave be one of depression, the motion is simply reversed in every respect.

If we have a series of waves of alternate elevation and depression, the elevations and depressions being equal in every respect, the motion of a surface-particle takes place in a closed curve, a complete revolution being performed during the transit of an elevation and the following depression. Compare Art. 157.

The horizontal motion is, in all cases, the same for particles in the same vertical; the vertical motion is everywhere small, and may (since it is zero at the bottom), be taken as proportional to the undisturbed height of the particle above the bed. The motion of a particle at a height y from the bottom may therefore be obtained from that of a surface-particle, by merely reducing the vertical component in the ratio $y : h$.

. 150. We can now examine under what circumstances the assumptions of Art. 147 are justified. The time which a wave (of elevation, say), of length λ takes to pass any particle is $\frac{\lambda}{c}$, so that if the slope of the wave-profile be gradual, the vertical velocity of a surface-particle is of the order $k \div \frac{\lambda}{c}$, where k is the maximum elevation above the undisturbed level. The horizontal velocity is, by (10), of the order $\frac{ck}{h}$; the ratio of the vertical to the horizontal velocity is therefore of the order $\frac{h}{\lambda}$. The assumption made at the beginning of Art. 147 requires then that the slope of the wave-profile should be gradual, and that the length of an elevation or depression should be large compared with the undisturbed depth of the fluid. Waves which fulfil these conditions are called 'long waves.' The motion of waves for which the conditions are not satisfied is discussed below, Art. 156.

The second assumption of Art. 147, viz. that $\frac{d\xi}{dx}$ is small, re-quires, by (4), that the maximum height of the wave should be small compared with the depth of the fluid.

151.   The potential energy of a wave, or system of waves, due to the elevation or depression of the fluid above the mean level h is, per unit breadth, $g\rho\iint y\,dx\,dy$, where the integration with respect to y is to be taken between the limits 0 and η, and that with respect to x over the whole length of the waves. Performing the former integration, we get

$$\tfrac{1}{2}g\rho \int \eta^2 dx \dots\dots\dots\dots\dots\dots\dots(11).$$

The kinetic energy is, in the case of Art. 147,

$$\tfrac{1}{2}\rho h \int \left(\frac{d\xi}{dt}\right)^2 dx \dots\dots\dots\dots\dots\dots\dots(12).$$

In a system of waves travelling in one direction only we have

$$\frac{d\xi}{dt} = \pm \frac{c}{h}\,\eta,$$

so that (11) and (12) are equal; or the total energy is half poten-tial, and half kinetic.

152.   If in any case of waves travelling in one direction only, without change of form, we impress on the whole mass a velocity equal and opposite to that of propagation, the motion becomes *steady*, whilst the forces acting on any particle remain the same as before. With the help of this artifice, the laws of wave-propa-gation can be investigated with great ease[*]. Thus, in the case of Art. 147 we have, at the free surface,

$$\frac{p}{\rho} = \text{const.} - \tfrac{1}{2}q^2 - g(h + \eta)\dots\dots\dots\dots\dots(13),$$

---

[*] See Lord Rayleigh, "On Waves," *Phil. Mag.*, April, 1876.

where q is the velocity. If the slope of the wave-profile be everywhere gradual, and the depth h small compared with the length of a wave, the horizontal velocity may be taken as uniform throughout the depth, and approximately equal to q. Hence the equation of continuity is

$$q(h + \eta) = ch,$$

c being the velocity, in the steady motion, where the depth is uniform and equal to h. Substituting for q in (13), we have

$$\frac{p}{\rho} = \text{const.} - \tfrac{1}{2}c^2\left(1 + \frac{\eta}{h}\right)^{-2} - gh\left(1 + \frac{\eta}{h}\right).$$

If $\frac{\eta}{h}$ be small, the condition for a free surface, viz. $p = \text{const.}$, is satisfied approximately, provided

$$c^2 = gh,$$

which is identical with equation (7).

If we take account of the second power of $\frac{\eta}{h}$ we find that at a part of the stream where the average elevation is k the condition for a free surface is better satisfied by

$$c^2 = gh\left(1 + \frac{3k}{h}\right),$$

or

$$c = \sqrt{gh}\left(1 + \tfrac{3}{2}\frac{k}{h}\right) \quad\dots\dots\dots\dots\dots\dots(14).$$

The higher portions of a wave therefore advance faster than the lower, so that the form of a wave continually changes as it proceeds. Thus, in the case of a wave of elevation only, the slope becomes gradually steeper in front, and more gentle behind until finally the conditions on which our investigations are based fail altogether to hold. The formula (14) seems due to Airy[*]. It is otherwise obvious from (4) that the accuracy of the equation

$$\frac{d^2\xi}{dt^2} = c^2\frac{d^2\xi}{dx^2},$$

---

[*] *Encyclopædia Metropolitana*, Vol. v., " Tides and Waves," Art. 208.

as a representative of (3) at any particular point of the canal is improved by making

$$c^2 = gh \left(1 - \frac{k}{h}\right)^{-3},$$

where k is the elevation in the neighbourhood of that point. If we neglect the square of $\frac{k}{h}$, this will agree with (14).

153. We proceed to investigate, on the same assumptions as before, the equations of motion of long waves in a uniform canal when, in addition to gravity, small disturbing forces whose horizontal and vertical components are X and Y act on the fluid. We have in this case,

$$\frac{p}{\rho} = g(h + \eta - y) - \int_{y}^{h+\eta} Y dy + \text{const.},$$

so that the equation of horizontal motion of a particle in the position (x, y), viz.

$$\rho \frac{d^2\xi}{dt^2} = -\frac{dp}{dx} + \rho X,$$

becomes

$$\frac{d^2\xi}{dt^2} = -g \frac{d\eta}{dx} + \int_{y}^{h+\eta} \frac{dY}{dx} dy + \frac{d\eta}{dx} Y + X.$$

If X and Y be of the same order, and their rates of variation small, the second and third terms on the right-hand side of this equation may be neglected in comparison with the other two. We then have

$$\frac{d^2\xi}{dt^2} = -g \frac{d\eta}{dx} + X \dots\dots\dots\dots\dots(15),$$

where, for the same reason, X may be supposed a function of x only. This equation being independent of y, the particles which at any instant lie in a plane perpendicular to x lie always in such a plane. The equation of continuity (4) then applies; and (15) becomes

$$\frac{d^2\xi}{dt^2} = c^2 \frac{d^2\xi}{dx^2} + X \dots\dots\dots\dots\dots(16),$$

where $c^2 = gh$, as before.

Since (16) does not contain Y, it appears that the horizontal disturbing force is much more effective than the vertical one in producing waves. This might be anticipated from the hydro-statical theorem that a liquid is in equilibrium when, and only when, the surfaces of equal pressure coincide with those of equal potential. The latter being everywhere perpendicular to the lines of force, it is plain that the addition of a small horizontal force would make them deviate from their original horizontal arrange-ment far more than the addition of a small vertical force.

154. We will follow Airy* in applying equation (16) to illus-trate the theory of the tides. Let us investigate the tides which would be produced in a uniform equatorial canal, the moon being supposed to describe a circle in the plane of the equator. Let x denote the undisturbed distance, measured along the equator, of a particle of water from some fixed meridian, $x + \xi$ the value of the same quantity, for the same particle, at the time t. If n be the angular velocity of the earth's rotation, the actual displace-ment of the particle at the time t is $\xi + nt$; so that the tangential acceleration will be $\dfrac{d^2\xi}{dt^2}$. If we suppose the 'centrifugal force' to be as usual allowed for in the value of g, the processes of the previous articles will apply without further alteration. Also, we have†, approximately

$$X = \mu \sin 2\theta,$$

where

$$\mu = \tfrac{3}{2} \frac{Ma}{D^3},$$

$M =$ the mass of the moon in astronomical units,

$a =$ the radius of the earth,

$D =$ the distance between the centres of the earth and the moon,

$\theta =$ the hour-angle of the moon at the station x,

$$= pt - \frac{x}{a} + a,$$

* Sect. VI., "Tides and Waves."
† Thomson and Tait, Art. 804.

provided $p = n + n'$, n' being the angular velocity of the moon in her orbit, and a some constant. The equation (16) then becomes

$$\frac{d^2\xi}{dt^2} = c^2 \frac{d^2\xi}{dx^2} + \mu \sin 2\left(pt - \frac{x}{a} + \alpha\right) \quad\dots\dots\dots(17).$$

The solution of this equation consists of two parts. The first part, or 'complementary function,' is the solution of (17) with the last term omitted, and expresses the free waves which could exist independently of the moon's action. The second part of the solution, or 'particular integral,' gives the forced waves or tides produced by the moon, and is

$$\xi = \tfrac{1}{4} \frac{\mu a^2}{c^2 - p^2 a^2} \sin 2\left(pt - \frac{x}{a} + \alpha\right).$$

The corresponding elevation η is given by

$$\eta = -h\frac{d\xi}{dx} = \tfrac{1}{2} \frac{\mu ah}{c^2 - p^2 a^2} \cos 2\left(pt - \frac{x}{a} + \alpha\right) \quad\dots\dots\dots(18).$$

The tide is therefore semidiurnal; and is 'direct' or 'inverted,' i.e. there is high or low water beneath the moon, according as c is greater or less than pa. Now

$$\frac{c^2}{p^2 a^2} = \frac{g}{n^2 a}\frac{h}{a}, \text{ nearly, } = 289\frac{h}{a},$$

in the actual case of the earth. Unless therefore the depth of the canal were much greater than the actual depth of the sea, the tides would be inverted.

For the case of a circular canal parallel to the equator in latitude λ, we should find

$$X = \mu \cos \lambda \sin 2\theta,$$

where

$$\theta = pt - \frac{x}{a \cos \lambda} + \alpha.$$

Substituting in (16), and solving as before, we find for the forced waves

$$\eta = \tfrac{1}{2} \frac{\mu ah \cos^2\lambda}{c^2 - p^2 a^2 \cos^2\lambda} \cos 2\left(pt - \frac{x}{a \cos \lambda} + \alpha\right) \quad\dots\dots(19).$$

If the latitude of the canal be higher than arc $\cos \dfrac{c}{pa}$, the tides will be direct.

155.   Next let us take the case of a canal coinciding with a meridian.   Let θ denote the hour-angle of the moon from this meridian, $= pt + a$, say, and x the distance of any point on the canal from the equator.   By an easy application of Spherical Trigonometry, we find for the horizontal disturbing force in the direction of the length of the canal,

$$X = - \mu \sin 2 \frac{x}{a} \cos^2 \theta$$

$$= - \tfrac{1}{2}\mu \sin 2 \frac{x}{a} \{1 + \cos 2 (pt + a)\}.$$

The equation of motion is easily seen to be of the same form, (16), as before.   Substituting then and solving, we find

$$\eta = \tfrac{1}{4} \frac{\mu a h}{c^2} \cos 2 \frac{x}{a} + \tfrac{1}{4} \frac{\mu a h}{c^2 - p^2 a^2} \cos 2 \frac{x}{a} \cos 2 (pt + a) \ldots \ldots (20).$$

The first term represents a permanent deviation of the surface from the circular form; the equation of the mean level being now

$$\eta = \tfrac{1}{4} \frac{\mu a h}{c^2} \cos 2 \frac{x}{a}.$$

The fluctuations above and below this mean level are given by the second term of (20).   If, as in the actual case of the earth, c be less than pa, there will be high water in latitudes above $45°$, and low water in latitudes below $45°$, when the moon is in the meridian of the canal, and *vice versa* when the moon is $90°$ from that meridian.   The circumstances are all reversed when c is greater than pa.

For a further development of the canal theory of the tides, the student is referred to Airy, *l. c. ante.*

Waves in deep water.

136.   When we abandon the assumption that the depth h of the fluid is small compared with the length of a wave, the Eulerian method becomes more appropriate.

Let the origin be in the undisturbed surface, the axis of x horizontal, that of y vertical and its positive direction upwards. We suppose the motion to take place entirely in these two dimensions x, y; and to be such as may have been generated from rest, so that there exists a single-valued velocity-potential ϕ. We retain, for a first approximation, the assumption that the squares and products of the velocities and relative displacements may be neglected. Our equations then are

$$\frac{d^2\phi}{dx^2} + \frac{d^2\phi}{dy^2} = 0 \dots\dots\dots\dots\dots\dots(21),$$

$$\frac{p}{\rho} = F(t) - gy - \frac{d\phi}{dt} \dots\dots\dots\dots\dots(22),$$

with the boundary conditions

$$\frac{dp}{dt} + \frac{d\phi}{dx}\frac{dp}{dx} + \frac{d\phi}{dy}\frac{dp}{dy} = 0 \dots\dots\dots\dots\dots(23)$$

at the free surface [(10) Art. 10,], and

$$\frac{d\phi}{dy} = 0 \dots\dots\dots\dots\dots\dots(24),$$

when $y = -h$.

Now (21) is satisfied by the sum of any number of terms of the form

$$e^{jx+ky},$$

each multiplied by an arbitrary function of t, provided $j^2 + k^2 = 0$. Now j must in our case be wholly imaginary, for otherwise we should have $\dfrac{d\phi}{dx}$ infinite for either $x = +\infty$, or $x = -\infty$. Hence we must have k real and $j = \pm ik$. We write therefore

$$\phi = \Sigma \{e^{ky}(A\cos kx + B\sin kx) + e^{-ky}(A'\cos kx + B'\sin kx)\},$$

the coefficients being functions of t as yet undetermined. The condition (24) gives

$$A e^{-kh} = A' e^{kh}, \qquad B e^{-kh} = B' e^{kh},$$

so that the assumed value of ϕ takes the form

$$\phi = \Sigma \{e^{k(y+h)} + e^{-k(y+h)}\}(P\cos kx + Q\sin kx)\dots\dots(25).$$

So far our work is rigorous. If we now neglect squares and products of small quantities, (23) becomes on substitution from (22)

$$F'(t) - \frac{d^2\phi}{dt^2} - g\frac{d\phi}{dy} = 0 \dots\dots\dots\dots\dots(26),$$

and in this we may suppose y equated to zero, for the error in the value of y will only introduce an error of the second order in this condition. Substituting the value (25) of ϕ, we find that in order that (26) may hold for all values of x we must have $F'(t) = 0$, and

$$(e^{kh} + e^{-kh})\frac{d^2 P}{dt^2} + gk(e^{kh} - e^{-kh})P = 0 \ldots\ldots\ldots(27),$$

with an equation of the same form for Q. The solution of (27) is

$$P = A \cos kct + B \sin kct,$$

where

$$c^2 = \frac{g}{k}\frac{e^{kh} - e^{-kh}}{e^{kh} + e^{-kh}} \ldots\ldots\ldots\ldots\ldots\ldots(28),$$

and A, B now denote absolute constants. Combining our results we get for ϕ a series of terms of the form

$$\alpha\left\{e^{k(y+h)} + e^{-k(y+h)}\right\}{\textstyle{\cos \atop \sin}} k(x \pm ct)\ldots\ldots \ldots(29),$$

where α is a constant.

Let us examine the motion represented by one of these terms alone, taking, say, in the last factor the cosine with the minus sign in its argument. The form of the free surface ($p = $ const.) is given by (22), viz. it is

$$y = \text{const.} - \frac{1}{g}\frac{d\phi}{dt},$$

where in the last term we suppose y equated to zero, for the reason already given. Hence if the origin be taken at the mean level, the equation of the free surface is

$$y = a \sin k(x - ct)\ldots\ldots\ldots\ldots\ldots\ldots(30),$$

where

$$a = -\frac{\alpha kc}{g}(e^{kh} + e^{-kh})\ldots\ldots\ldots\ldots\ldots(31).$$

The wave-profile is therefore the curve of sines, and it advances without change of form in the direction of x positive, with uniform velocity c given by (28). The 'wave-length,' i.e. the distance between two successive crests or hollows is

$$\lambda = \frac{2\pi}{k}.$$

It appears that the velocity of propagation is not independent of the wave-length, but that it increases continuously with the value of the ratio $\frac{h}{\lambda}$, being $\sqrt{gh}$ (as in Art. 148) when this ratio is infinitesimal, and $\sqrt{\frac{g\lambda}{2\pi}}$ when it is infinite. To any given value of c there corresponds only one value of λ, and *vice versa*.

157.   Let us examine the nature of the motion of the individual particles of fluid as a system of waves of the above kind passes over them. If ξ, η be the component displacements at time t of the particle whose mean position is (x, y), we have

$$\frac{d\xi}{dt} = \frac{d\phi}{dx} = -ka\{e^{k(y+h)} + e^{-k(y+h)}\} \sin k(x-ct),$$

$$\frac{d\eta}{dt} = \frac{d\phi}{dy} = \quad ka\{e^{k(y+h)} - e^{-k(y+h)}\} \cos k(x-ct).$$

We ought, in strictness, to have on the right-hand side of these equations $x + \xi$ for x, and $y + \eta$ for y, but the resulting correction would be of the order a^2 which we have agreed to neglect. Integrating then the above equations on the supposition that x, y are constant, and remembering the formulæ (28), (31) for c and a, we find

$$\left. \begin{array}{l} \xi = a\,\dfrac{e^{k(y+h)} + e^{-k(y+h)}}{e^{kh} - e^{-kh}} \cos k(x-ct), \\[3mm] \eta = a\,\dfrac{e^{k(y+h)} - e^{-k(y+h)}}{e^{kh} - e^{-kh}} \sin k(x-ct). \end{array} \right\} \quad \ldots\ldots\ldots\ldots(32).$$

Each particle therefore describes an ellipse whose major axis is horizontal; the law of description being the same as for a particle attracted to a fixed point by a force varying as the distance. The ratio of the minor to the major axis of the ellipse, viz.

$$\frac{e^{k(y+h)} - e^{-k(y+h)}}{e^{k(y+h)} + e^{-k(y+h)}},$$

diminishes from the surface to the bottom, where it vanishes.

If we compare (32) with (30), we see that a surface-particle moves in the direction of the waves when it coincides with a crest, and in the opposite direction when it coincides with a trough.

If the ratio $\dfrac{h}{\lambda}$ be infinite, the expressions (32) become

$$\xi = ae^{ky} \cos k(x - ct), \qquad \eta = ae^{ky} \sin k(x - ct),$$

so that each particle describes a vertical circle with uniform angular velocity. The radii of these circles, given by the formula ae^{ky}, diminish rapidly with increasing depth*. At a depth below the surface equal to the wave-length the motion is to that of the surface in the ratio $e^{-2\pi} : 1$, or $1 : 535$ nearly.

158. The energy of a system of waves of the kind now under examination is found as follows. Suppose two vertical planes drawn perpendicular to the ridges of the waves, at unit distance apart. The potential energy of the fluid between these planes is $\frac{1}{2}g\rho\int y^2 dx$, where y is the elevation above the mean level at the point x. Substituting from (30), and integrating, we find $\frac{1}{4}g\rho a^2\lambda$ for the potential energy per wave-length.

The kinetic energy is

$$\tfrac{1}{2}\rho \iint \left\{ \left(\frac{d\phi}{dx}\right)^2 + \left(\frac{d\phi}{dy}\right)^2 \right\} dx\,dy$$

$$= \tfrac{1}{2}k^2a^2\rho \iint \{ e^{2k(y+h)} + e^{-2k(y+h)} + 2\cos 2k(x - ct) \} dx\,dy.$$

This gives when integrated with respect to x over a wave-length, and with respect to y between the limits $-h$ and 0,

$$\tfrac{1}{4}k a^2\rho\lambda \left(e^{2kh} - e^{-2kh} \right),$$

or, by (28) and (31), $\frac{1}{4}g\rho a^2\lambda$.

The kinetic and potential energies are therefore equal.

159. So long as we confine ourselves to a first approximation all our equations are linear; so that if ϕ_1, ϕ_2, &c. be the velocity-

* These results were given by Green, *Camb. Trans.* Vol. VII. 1839.

potentials of distinct systems of waves of the particular kind above considered, then

$$\phi = \phi_1 + \phi_2 + \&c.\quad\quad\quad\quad\quad\quad(33)$$

will be the velocity-potential of a possible form of wave-motion, with a free surface. Since when ϕ is determined the equation of the free surface is given by

$$y = -\frac{1}{g}\left[\frac{d\phi}{dt}\right]_{y=0},$$

the elevation above the mean level at any point of the surface, in the motion given by (33), will be equal to the algebraic sum of the elevations due to the separate systems of waves ϕ_1, ϕ_2, &c. Hence each of the latter systems is propagated exactly as if the others were absent, and produces its own elevation or depression at each point of the surface.

We may in this way by adding together terms of the form (29), with properly chosen values of α, build up the solution of the general problem of Art. 156 in the case where the initial conditions are any whatever. Thus, let us suppose that, when $t = 0$, the equation of the free surface is

$$y = f(x),$$

and that the normal velocity at that surface is then $F(x)$, or, to our order of approximation,

$$\left[\frac{d\phi}{dy}\right]_{y=0} = F(x).$$

The value of ϕ is found to be

$$\phi = \int_0^\infty \frac{dk}{\pi}\frac{g}{kc}\frac{e^{k(y+h)}+e^{-k(y+h)}}{e^{kh}+e^{-kh}}\left[\frac{1}{kc}\left\{\int_{-\infty}^\infty d\lambda\, F(\lambda)\cos k(\lambda - x)\right\}\cos kct\right.$$
$$\left. - \left\{\int_{-\infty}^\infty d\lambda f(\lambda)\cos k(\lambda - x)\right\}\sin kct\right],$$

and the equation of the free surface is

$$y = \int_0^\infty \frac{dk}{\pi}\left[\left\{\int_{-\infty}^\infty d\lambda f(\lambda)\cos k(\lambda - x)\right\}\cos kct\right.$$
$$\left. + \frac{1}{kc}\left\{\int_{-\infty}^\infty d\lambda F(\lambda)\cos k(\lambda - x)\right\}\sin kct\right].$$

These formulæ, in which c is a function of k given by (28), may be readily verified by means of Fourier's expression for an arbitrary function as a definite integral, viz.

$$f(x) = \frac{1}{\pi} \int_0^\infty dk \left\{ \int_{-\infty}^\infty d\lambda f(\lambda) \cos k (\lambda - x) \right\}.$$

When the initial conditions are arbitrary, the subsequent motion is made up of systems of waves, of all possible lengths, travelling in either direction, each with the velocity proper to its own wave-length, as given by (28). Hence, in general, the form of the free surface is continually altering, the only exception being when the wave-length of every component system which is present in sensible amplitude is large compared with the depth of the fluid. In this case the velocity of propagation $\sqrt{gh}$ is independent of the wave-length, so that if we have waves travelling in one direction only, the wave-profile remains un-changed in form as it advances. Compare Art. 148.

A curious result of the dependence of the velocity of propaga-tion on the wave-length occurs when we have two systems of waves of the same amplitude, and of nearly but not quite equal wave-lengths, travelling in the same direction. The equation of the free surface is then of the form

$$y = a \sin k (x - ct) + a \sin k' (x - c't)$$

$$= 2a \cos \left(\frac{k - k'}{2} x - \frac{kc - k'c'}{2} t \right) \sin \left(\frac{k + k'}{2} x - \frac{kc + k'c'}{2} t \right).$$

If k, k' be very nearly equal, the cosine in this expression varies very slowly with x and t; so that the wave-profile at any instant is in the form of a curve of sines in which the amplitude alternates slowly between the values 0 and $2a$. The surface therefore presents the appearance of a series of groups of waves separated at equal intervals by bands of nearly smooth water. The interval between the centres of two successive groups is $\dfrac{2\pi}{k - k'}$, and the velocity of advance of the *groups* is $\dfrac{kc - k'c'}{k - k'}$, or $\dfrac{d \cdot kc}{dk}$, ultimately.

From (28) we find

$$\frac{d \cdot kc}{dk} = \tfrac{1}{2}c \left(1 + \frac{4kh}{e^{2kh} - e^{-2kh}}\right).$$

The ratio of this velocity of propagation to that of the *waves* increases as kh diminishes, being $\tfrac{1}{2}$ when the depth is infinite, and unity when it is small compared with the wave-length*.

160. Rankine† has by a synthetic process arrived at the following *exact* equations, expressing a possible form of wave-motion when the depth of the fluid is infinite, viz.:

$$\left. \begin{aligned} x &= a + \frac{1}{k} e^{-kb} \sin k\,(a + ct) \\ y &= b + \frac{1}{k} e^{-kb} \cos k\,(a + ct) \end{aligned} \right\} \quad \ldots\ldots\ldots\ldots (34).$$

Here x, y are the co-ordinates at the time t of the particle defined by the parameters a, b, whilst k, c are absolute constants. The axis of x is horizontal, that of y vertical and downwards.

We find at once

$$\frac{d\,(x,\,y)}{d\,(a,\,b)} = 1 - e^{-2kb},$$

so that the Lagrangian equation of continuity [see Art. 17] is satisfied. Also, substituting from (34) in the Lagrangian equations of motion, we have

$$\frac{d}{da}\left(\frac{p}{\rho} + V\right) = kc^2\,e^{-kb} \sin k\,(a + ct),$$

$$\frac{d}{db}\left(\frac{p}{\rho} + V\right) = kc^2\,e^{-kb} \cos k\,(a + ct) - kc^2\,e^{-2kb},$$

whence, since $V = -gy$,

$$\frac{p}{\rho} = g\left\{b + \frac{1}{k} e^{-kb} \cos k\,(a + ct)\right\} - c^2 e^{-kb} \cos k\,(a + ct) + \tfrac{1}{2}c^2\,e^{-2kb}.$$

Now at the free surface p must be independent of t, so that

$$g = kc^2,$$

---

* Stokes, Smith's Prize Examination, 1876. See also Prof. O. Reynolds, ' On Waves,' *Nature*, Vol. xvi., p. 343; and Lord Rayleigh, *Theory of Sound*, Art. 191. and *Proc. Lond. Math. Soc.* Nov. 8, 1877.

† *Phil. Trans.* 1863.

and
$$\frac{p}{\rho} = gb + \tfrac{1}{2} c^2 e^{-2kb}.$$

The pressure about any particle is therefore constant during the motion. The form of any surface of equal pressure ($b = $ const.) is obtained by rolling a circle of radius $\frac{1}{k}$ on the under side of the straight line $y = b - \frac{1}{k}$. A point fixed relatively to this circle, at a distance $\frac{1}{k} e^{-kb}$ from its centre, will trace out the profile of the surface in question. The wave-length is

$$\lambda = \frac{2\pi}{k},$$

and the velocity of propagation is

$$c = \sqrt{\frac{g}{k}} = \sqrt{\frac{g\lambda}{2\pi}}.$$

If we differentiate (34) with respect to t, we get for the component velocities of a particle

$$u = ce^{-kb} \cos k\,(a + ct),$$
$$v = -ce^{-kb} \sin k\,(a + ct),$$

and thence

$$u\,dx + v\,dy = d\left\{\frac{c}{k} e^{-kb} \sin k\,(a + ct)\right\} + ce^{-2kb}\,da,$$

which is not an exact differential, so that the motion represented by (34) is rotational, and cannot therefore have been generated from rest by the action of ordinary natural forces. The circulation in the boundary of the parallelogram whose angular points coincide with the particles

$$(a,\ b),\ (a + da,\ b),\ (a,\ b + db),\ (a + da,\ b + db)$$

is
$$-\frac{d}{db}\left(ce^{-2kb}\right)\,da\,db,$$

and the area of this parallelogram is

$$\frac{d\,(x,\ y)}{d\,(a,\ b)}\,da\,db = \left(1 - e^{-2kb}\right)\,da\,db,$$

L.   13

so that the angular velocity of the element (a, b) is

$$\frac{kce^{-2kb}}{1 - e^{-2kb}}.$$

This is greatest at the surface, and diminishes rapidly with increasing depth.

Propagation in Two Dimensions.

161. In the cases already considered the propagation of the waves over the surface of the fluid has been supposed to take place in one dimension (x) only. We will now sketch the method to be pursued in treating cases where the propagation is in two dimensions.

Let the origin be taken in the undisturbed surface, the axes of x and y horizontal, that of z vertical and upwards; and let h be the (uniform) depth of the fluid. The velocity-potential ϕ must satisfy

$$\nabla^2\phi = 0 \dots\dots\dots\dots\dots\dots(35),$$

and the condition

$$\frac{d\phi}{dz} = 0, \quad \text{when } z = -h\dots\dots\dots\dots\dots(36).$$

Further, at the free surface we must have, making the same approximations as in Art. 156

$$\frac{dp}{dt} - g\frac{d\phi}{dz} = 0,$$

where

$$\frac{p}{\rho} = \text{const.} - \frac{d\phi}{dt} - gz,$$

or

$$\frac{d^2\phi}{dt^2} + g\frac{d\phi}{dz} = 0, \quad [z = 0]\dots\dots\dots\dots(37).$$

Now (35) is satisfied by the sum of any number of terms of the form

$$\{e^{k(z+h)} + e^{-k(z+h)}\}\,\phi' \dots\dots\dots\dots\dots(38),$$

where the k's are constants yet to be determined, and ϕ' does not contain z. Substituting in (37) we find

$$(e^{kh} + e^{-kh}) \frac{d^2\phi'}{dt^2} + gk(e^{kh} - e^{-kh}) \phi' = 0,$$

or $\qquad\qquad \phi' = \phi_1 \cos kct + \phi_2 \sin kct \dots\dots\dots\dots\dots(39),$

where c is given by

$$c^2 = \frac{g}{k} \frac{e^{kh} - e^{-kh}}{e^{kh} + e^{-kh}},$$

and ϕ_1, ϕ_2 are functions of x, y only, to be determined from (35) which now gives

$$\frac{d^2\phi_1}{dx^2} + \frac{d^2\phi_1}{dy^2} + k^2\phi_1 = 0\dots\dots\dots\dots\dots(40),$$

with a similar equation for ϕ_2. The solution of (40), when adapted to suit the conditions to be satisfied at the surfaces, if any, which limit the fluid horizontally, gives the possible values of k. By adding together terms of the form (38), in which k has the values thus found, we may build up a solution satisfying any arbitrary initial conditions.

Oscillations in a Rectangular Tank.

162. We apply the method just explained to the case where the fluid is contained in a rectangular tank whose sides are vertical. Let the origin be taken in one corner of the tank, and the axes of x, y along two of its sides, and let the equations of the other two sides be $x = a$, $y = b$, respectively. The functions ϕ_1, ϕ_2 must now satisfy the conditions

$$\frac{d\phi_1}{dx} = 0, \text{ when } x = 0, \text{ and when } x = a,$$

and

$$\frac{d\phi_1}{dy} = 0, \text{ when } y = 0, \text{ and when } y = b.$$

The general value of ϕ_1 subject to these conditions is given by the double Fourier's series

$$\phi_1 = \Sigma\Sigma A_{m,n} \cos \frac{m\pi x}{a} \cos \frac{n\pi y}{b},$$

13—2

where the summations include all integral values of m, n from 1 to ∞. Substituting in (40), we find

$$k^2 = \pi^2 \left(\frac{m^2}{a^2} + \frac{n^2}{b^2} \right).$$

If a be the greater of the two quantities a, b, the component oscillation of longest period is got by making $m = 1$, $n = 0$, whence

$$k = \frac{\pi}{a};$$

the motion is then parallel to the longer side of the tank, and consists of two systems of waves of the kind considered in Art. 156, travelling in opposite directions, the wave-length being $2a$.

Circular Tank.

163. In the case of a circular tank, whose axis is vertical, it is convenient to take the origin in the axis, and to transform to polar co-ordinates by writing

$$x = r \cos\theta, \qquad y = r \sin\theta.$$

The equation (40) then becomes

$$\frac{d^2\phi_1}{dr^2} + \frac{1}{r}\frac{d\phi_1}{dr} + \frac{1}{r^2}\frac{d^2\phi_1}{d\theta^2} + k^2\phi_1 = 0 \ldots\ldots\ldots(41).$$

Now whatever be the value of ϕ' it can be expanded by Fourier's theorem in the form

$$\phi' = \Sigma\,(\psi_n \cos n\theta + \chi_n \sin n\theta),$$

where the summation embraces all integral values of n, and ψ_n, χ_n are functions of r only. Substituting in (41), we have, to determine ψ_n,

$$\frac{d^2\psi_n}{dr^2} + \frac{1}{r}\frac{d\psi_n}{dr} + \left(k^2 - \frac{n^2}{r^2} \right)\psi_n = 0 \ldots\ldots\ldots\ldots(42),$$

with an equation of the same form for χ_n.

The solution of (42), subject to the condition that ψ_n is finite when $r = 0$, is

$$\psi_n = \Sigma A_n J_n(kr) \ldots\ldots\ldots\ldots\ldots\ldots\ldots(43),$$

where A_n is an arbitrary constant, and $J_n(kr)$ denotes the Bessel's Function of the nth order of the variable kr, viz.*

$$J_n(kr) = \frac{(kr)^n}{2^n \lfloor n} \left\{ 1 - \frac{(kr)^2}{2(2n+2)} + \frac{(kr)^4}{2.4(2n+2)(2n+4)} - \&c. \right\}$$

The summation in (43) is supposed to include all admissible values of k, to be determined by the equation

$$\frac{d\psi_n}{dr} = 0, \text{ when } r = a,$$

or $$J_n'(ka) = 0 \dots \dots (44),$$

the accent denoting the first derived function, and a being the radius of the tank. It may be shewn that the values of k satisfying (44) are infinite in number and all real. For the particular case $n = 0$, when the motion is symmetrical about the centre, the lowest roots are given by

$$\frac{ka}{\pi} = 1\cdot2197, \ 2\cdot2330, \ 3\cdot2383, \ \&c.$$

For a discussion of the various kinds of motion represented by the above formulæ the reader is referred to Lord Rayleigh's paper on Waves already cited†, which contains besides a comparison with theory of some experimental measurements of the periods of oscillation in rectangular and circular tanks made by Guthrie‡.

Free oscillations of an Ocean of uniform Depth.

164. We close this chapter with the discussion of the following problem, which is of some interest in connexion with the theory of the tides :—To determine the free oscillations of an ocean of uniform depth completely enveloping a spherical earth. The method employed is due to Thomson§.

Let r denote the distance of any point from the centre of the sphere, and a, b the values of r at the surface of the solid sphere, and at the mean level of the ocean, respectively. The general

---

* Todhunter, *Functions of Laplace, &c.*, Art. 370.

† See also *Theory of Sound*, c. 9, by the same author. The values of the roots of (44) for the case $n = 0$ are taken from this source.

‡ *Phil. Mag.* 1875.

§ *Phil. Trans.* 1863, p. 608.

solution of the equation of continuity $\nabla^2\phi = 0$, subject to the condition that $\dfrac{d\phi}{dr} = 0$ when $r = a$, is

$$\phi = \Sigma \left\{ (n+1)\left(\frac{r}{a}\right)^n + n\left(\frac{a}{r}\right)^{n+1} \right\} S_n \quad\dots\dots\dots\dots(45),$$

where S_n is the general surface-harmonic of order n. The $2n + 1$ arbitrary constants which the general expression for S_n contains are functions of t, to be determined. The formula for the pressure is, if we neglect squares and products of small quantities,

$$\frac{p}{\rho} = C - \frac{d\phi}{dt} + V \quad\dots\dots\dots\dots\dots(46),$$

where ρ is the density of the fluid, V the potential due to the joint attraction of the earth and the sea. If we assume as the equation of the free surface at time t

$$r = b + \Sigma T_n \quad\dots\dots\dots\dots\dots\dots(47),$$

where T_n is a spherical harmonic of order n, we have*, at this surface,

$$V = \frac{E}{r} + 4\pi\rho b\, \Sigma\, \frac{T_n}{2n+1}.$$

Here E denotes the total mass of the earth and sea, viz.

$$E = \tfrac{4}{3}\pi a^3\sigma + \tfrac{4}{3}\pi(b^3 - a^3)\rho,$$

if σ be the mean density of the earth. Substituting in (46), expanding, and omitting as before the squares of small quantities we find, at the free surface,

$$\frac{p}{\rho} = C - \Sigma \left\{ (n+1)\left(\frac{b}{a}\right)^n + n\left(\frac{a}{b}\right)^{n+1} \right\} \frac{dS_n}{dt} - \frac{E}{b^2}\Sigma T_n + 4\pi\rho b\, \Sigma\, \frac{T_n}{2n+1}.$$

But at the free surface $p = \text{const.}$ If we put $g = \dfrac{E}{b^2}$, this consideration gives

$$g\left\{ 1 - \frac{3}{2n+1}\, \frac{\rho b^3}{\sigma a^3 + \rho(b^3 - a^3)} \right\} T_n$$

$$= -\left\{ (n+1)\left(\frac{b}{a}\right)^n + n\left(\frac{a}{b}\right)^{n+1} \right\} \frac{dS_n}{dt} \quad\dots\dots\dots(48).$$

---

* Pratt, *Figure of the Earth*, c. 3. It is assumed of course that E, ρ, σ are all expressed in 'astronomical units.'

We have still to express the condition that (47) should be the equation of a bounding surface. This condition [(10) Art. 10,] becomes in the present case

$$\Sigma \frac{dT_n}{dt} = \frac{d\phi}{dr},$$

whence

$$\frac{dT_n}{dt} = \frac{n(n+1)}{a} \left\{ \left(\frac{b}{a}\right)^{n-1} - \left(\frac{a}{b}\right)^{n+2} \right\} S_n \dots\dots\dots (49).$$

Eliminating S_n between (48) and (49) we find

$$\frac{d^2 T_n}{dt^2} + \frac{4\pi^2}{\tau_n^2} T_n = 0,$$

provided

$$\tau_n^2 = \frac{4\pi^2 a}{g} \left\{ (n+1)\left(\frac{b}{a}\right)^n + n\left(\frac{a}{b}\right)^{n+1} \right\}$$

$$\div n(n+1) \left\{ \left(\frac{b}{a}\right)^{n-1} - \left(\frac{a}{b}\right)^{n+2} \right\} \left\{ 1 - \frac{3}{2n+1} \frac{\rho b^3}{\sigma a^3 + \rho(b^3 - a^3)} \right\}$$
$$\dots\dots\dots(50).$$

The motion consists therefore in general of a series of superposed oscillations, the periods τ of which are obtained by putting $n = 1$, 2, 3, &c., in (50).

The longest period is that for which $n = 1$, in which case

$$\tau_1^2 = \frac{2\pi^2 b}{g} \frac{2b^3 + a^3}{b^3 - a^3} \div \left\{ 1 - \frac{\rho b^3}{\sigma a^3 + \rho(b^3 - a^3)} \right\} \dots\dots(51).$$

If $\sigma = \rho$, we have $\tau_1 = \infty$, as we should expect. In fact

$$r = b + T_1 \dots\dots\dots\dots\dots\dots\dots(52).$$

(T_1 infinitely small) is the equation of a sphere of radius b whose centre is near the origin. The fluid and solid are then equivalent to a single spherical mass of uniform density, so that there is always equilibrium when the surface is of the form (52).

If $\rho > \sigma$, τ_1 is imaginary, the value of τ_1^2 given by (51) being then negative. This indicates that the equilibrium of the ocean, when in the form of a sphere concentric with the earth, is unstable. The ocean would in fact, if disturbed, tend to heap itself up on one side*.

---

* See Thomson and Tait, Nat. Phil. Art. 816.

If $a = 0$, we get the case treated by Thomson, viz. the case in which a mass of liquid oscillates about the spherical form under the mutual gravitation of its parts. The formula (50) becomes

$$\tau_n^2 = \frac{4\pi^2 b}{g} \frac{2n+1}{2n(n-1)}.$$

"It is worthy of remark that the period of vibration thus calculated is the same for the same density of liquid, whatever be the dimensions of the globe.

"For the case of $n = 2$, or an ellipsoidal deformation, the length of the isochronous simple pendulum becomes $\frac{5}{4}b$, or one and a quarter times the earth's radius, for a homogeneous liquid globe of the same mass and diameter as the earth; and therefore for this case, or for any homogeneous liquid globe of about $5\frac{1}{2}$ times the density of water, the half-period is $47^m 12^s*$."

"A steel globe of the same dimensions, without mutual gravitation of its parts, could scarcely oscillate so rapidly, since the velocity of plane waves of distortion in steel is only about 10,140 feet per second, at which rate a space equal to the earth's diameter would not be travelled in less than $1^h 8^m 40^s†$."

If on the other hand the depth h of the ocean be small compared with the radius of the earth, we have, writing in (50) $b = a + h$, and neglecting squares, &c. of $\dfrac{h}{a}$,

$$\tau^2 = \frac{4\pi^2 a^2}{gh} \div n(n+1)\left(1 - \frac{3}{2n+1}\frac{\rho}{\sigma}\right) \quad \ldots\ldots\ldots(53),$$

a result due to Laplace‡.

For large values of n the distance from crest to crest in the surface represented by (47) is small compared with the radius of the earth. The propagation of the disturbance then takes place according to the laws investigated in Art. 147. The formula (53) then becomes, approximately,

$$\tau = 2\pi a \div n\sqrt{gh},$$

a result which the student who is familiar with the properties of spherical harmonics will easily see to be consistent with (7).

* *Phil. Trans.* 1863, p. 610.
† *Phil. Trans.* 1863, p. 573.
‡ *Mécanique Céleste*, Livre 1me, Art. 1.

CHAPTER VIII.

WAVES IN AIR [*].

165. We investigate in this chapter the general laws of the propagation of small disturbances produced in a mass of air (or other gas) originally at rest at a uniform temperature, avoiding such details as are more properly treated in books specially devoted to the theory of Sound.

Plane waves.

We take first the case where the motion is in one dimension x only. Let ξ be the displacement at time t of the particles which in the undisturbed state occupy the position x. The stratum of air originally bounded by the planes x and $x + dx$ is at the time t bounded by the planes $x + \xi$, and $x + \xi + \left(1 + \dfrac{d\xi}{dx}\right) dx$, so that the equation of continuity is

$$\rho\left(1 + \frac{d\xi}{dx}\right) = \rho_0 \dotfill (1),$$

where ρ_0 is the density in the undisturbed state. The equation of motion of the stratum is

$$\rho_0 \frac{d^2\xi}{dt^2} = -\frac{dp}{dx} \dotfill (2),$$

and if we suppose the condensations and rarefactions to succeed one another so rapidly that there is no sensible gain or loss of heat in any stratum by conduction or radiation, the relation between p and ρ is

$$p = k'\rho^\gamma \dotfill (3).$$

* This chapter was written independently of the corresponding portion of Lord Rayleigh's *Theory of Sound*, the second volume of which did not come into the author's hands until after the MS. of this treatise had been despatched to England (October 1878).

Eliminating p, ρ we find

$$\frac{d^2\xi}{dt^2} = c^2 \frac{\dfrac{d^2\xi}{dx^2}}{\left(1 + \dfrac{d\xi}{dx}\right)^{\gamma+1}} \dotfill (4),$$

where

$$c^2 = k'\gamma\rho_0{}^{\gamma-1} = \left[\frac{dp}{d\rho}\right]_{\rho=\rho_0} = \gamma\frac{p_0}{\rho_0} \dotfill (5).$$

166. Let us denote by s the 'condensation,' *i.e.* the value of $\dfrac{\rho - \rho_0}{\rho_0}$ at any point. If this be small, we have, by (1),

$$s = -\frac{d\xi}{dx},$$

nearly; and if as in all ordinary cases of sound the condensation, and its rate of variation from point to point, be both small quantities whose squares, &c., may be neglected, the equation (4) becomes

$$\frac{d^2\xi}{dt^2} = c^2 \frac{d^2\xi}{dx^2} \dotfill (6).$$

The complete solution of this is

$$\xi = F(x - ct) + f(x + ct) \dotfill (7),$$

which denotes two systems of waves travelling with velocity c, one in the positive, the other in the negative direction of x. Comparing (5) with the ordinary expression of the laws of Boyle and Charles, viz.

$$p = k\rho(1 + \alpha\theta),$$

we find

$$c^2 = k\gamma(1 + \alpha\theta),$$

so that the velocity of propagation c depends, for the same gas, only on the temperature θ.

For a wave travelling in one direction only, say that of x positive, we have

$$u = \frac{d\xi}{dt} = -cF'(x - ct),$$

$$s = -\frac{d\xi}{dx} = -F'(x - ct),$$

and therefore

$$u = cs \dots\dots\dots\dots\dots\dots\dots(8).$$

For a wave travelling in the direction of x negative we should have

$$u = -cs \dots\dots\dots\dots\dots\dots\dots(9).$$

It will be noticed that there is an exact correspondence between the above approximate theory, and that of 'long' waves in water (Art. 147). By the substitution of the words 'condensation of air' for 'elevation of water,' and 'rarefaction' for 'depression,' the two questions become identical.

The analogy becomes however less close when we proceed to a higher degree of approximation.

Spherical waves.

167. Let us suppose that the disturbance is symmetrical with respect to a fixed point, which we take as origin. The motion is then necessarily irrotational, so that a velocity-potential ϕ exists, which is a function of r, the distance of any point from the origin, and t, only. If as before we neglect the squares of small quantities, we have

$$\int \frac{dp}{\rho} = -\frac{d\phi}{dt}.$$

Now

$$\int \frac{dp}{\rho} = k' \frac{\gamma}{\gamma-1} (\rho^{\gamma-1} - \rho_0^{\gamma-1}),$$

whence, writing $\rho = \rho_0(1+s)$, and neglecting the square of s, we find

$$c^2 s = -\frac{d\phi}{dt} \dots\dots\dots\dots\dots\dots(10),$$

where c has the same value as in (5).

To form the equation of continuity we remark that, owing to the difference of flux across its inner and outer surfaces, the space bounded by the spheres r and $r + dr$ is gaining matter at the rate

$$-\frac{d}{dr} \left(4\pi r^2 \rho \frac{d\phi}{dr} \right) dr.$$

The same rate being also expressed by

$$4\pi r^2 dr \frac{d\rho}{dt},$$

we have

$$\frac{d\rho}{dt} + \frac{1}{r^2}\frac{d}{dr}\left(r^2\rho\frac{d\phi}{dr}\right) = 0 \dots \dots \dots (11),$$

a result which might also have been arrived at by direct transformation of the general equation of continuity. To our order of approximation, (11) gives .

$$\frac{ds}{dt} + \frac{1}{r^2}\frac{d}{dr}\left(r^2\frac{d\phi}{dr}\right) = 0,$$

and eliminating s between this and (10), we find

$$\frac{d^2\phi}{dt^2} = \frac{c^2}{r^2}\frac{d}{dr}\left(r^2\frac{d\phi}{dr}\right),$$

or

$$\frac{d^2 \cdot r\phi}{dt^2} = c^2\frac{d^2 \cdot r\phi}{dr^2} \dots \dots \dots \dots \dots (12).$$

This is of the same form as (6), so that the solution is

$$r\phi = F(r - ct) + f(r + ct) \dots \dots \dots \dots (13).$$

Hence the motion is made up of two systems of spherical waves travelling, one outwards, the other inwards, with velocity c. Considering for a moment the first system alone, we have

$$s = \frac{c}{r} F'(r - ct),$$

which shews that a condensation is propagated outwards with velocity c, but diminishes as it proceeds, its amount varying inversely as the distance from the origin. The velocity due to the same train of waves is

$$\frac{d\phi}{dr} = \frac{1}{r} F'(r - ct) - \frac{1}{r^2} F(r - ct).$$

As r increases the second term becomes less and less important compared with the first, so that ultimately the velocity is propagated according to the same law as the condensation.

168. Let us suppose that the initial distributions of velocity and condensation are given by the formulæ

$$\phi = \psi(r), \quad \frac{d\phi}{dt} = \chi(r) \dots \dots \dots \dots (14),$$

where ψ, χ are any arbitrary functions. The value of ϕ at any subsequent time t is then to be found as follows. Comparing (14) with (13) we find

$$F(r) + f(r) = r\psi(r),$$
$$-F'(r) + f'(r) = \frac{r}{c}\chi(r) \qquad \Bigg\} \dots\dots\dots\dots(15).$$

Integrating the latter equation, and putting $\int_0^r r\chi(r)\,dr = \chi_1(r)$, we obtain

$$-F(r) + f(r) = \frac{1}{c}\chi_1(r) + C,$$

where C is an arbitrary constant. This gives, in conjunction with (15),

$$F(r) = \tfrac{1}{2}r\psi(r) - \tfrac{1}{2}\frac{1}{c}\chi_1(r) - \tfrac{1}{2}C\dots\dots\dots\dots(16),$$

$$f(r) = \tfrac{1}{2}r\psi(r) + \tfrac{1}{2}\frac{1}{c}\chi_1(r) + \tfrac{1}{2}C\dots\dots\dots\dots(17).$$

The complete value (13) of ϕ is then found by writing, for r, $r - ct$ in (16), and $r + ct$ in (17), viz.

$$r\phi = \tfrac{1}{2}(r - ct)\psi(r - ct) - \tfrac{1}{2}\frac{1}{c}\chi_1(r - ct)$$
$$+ \tfrac{1}{2}(r + ct)\psi(r + ct) + \tfrac{1}{2}\frac{1}{c}\chi_1(r + ct)\dots\dots\dots(18).$$

It is obvious from the symmetry of the motion with respect to the origin that

$$\psi(-r) = \psi(r), \qquad \chi(-r) = \chi(r)\dots\dots\dots\dots(19),$$

and therefore

$$\psi'(-r) = -\psi'(r), \qquad \chi_1'(-r) = -\chi_1'(r)\dots\dots\dots\dots(20).$$

We shall require shortly the value of ϕ at the origin. This may be found by dividing both sides of (18) by r, and evaluating the indeterminate form which the right-hand member assumes for $r = 0$; or more simply by differentiating both sides of (18) with respect to r and then making $r = 0$. The result is, if we take account of the relations (19) and (20),

$$[\phi]_{r=0} = \frac{d}{dt}\cdot t\psi(ct) + t\chi(ct)\dots\dots\dots\dots(21).$$

169. We proceed to the general case of propagation of air-waves. We neglect, as before, the squares of small quantities, so that the dynamical equation is

$$c^2 s = -\frac{d\phi}{dt} \quad \dots\dots\dots\dots\dots\dots\dots\dots(10).$$

Also, writing $\rho = \rho_0(1 + s)$ in the general equation of continuity, (8) of Art. 8, we have, to the same order of approximation

$$\frac{ds}{dt} + \frac{d^2\phi}{dx^2} + \frac{d^2\phi}{dy^2} + \frac{d^2\phi}{dz^2} = 0 \dots\dots\dots\dots\dots(22).$$

The elimination of s between (10) and (22) gives

$$\frac{d^2\phi}{dt^2} = c^2 \left(\frac{d^2\phi}{dx^2} + \frac{d^2\phi}{dy^2} + \frac{d^2\phi}{dz^2} \right),$$

or, with our former notation,

$$\frac{d^2\phi}{dt^2} = c^2 \nabla^2 \phi \dots\dots\dots\dots\dots\dots\dots(23).$$

170. Let us suppose a sphere of radius r described about any point (x, y, z) as centre. Multiplying both sides of (23) by $dx\,dy\,dz$, and integrating throughout the volume of the sphere, we find

$$\frac{d^2}{dt^2} \iiint \phi\,dx\,dy\,dz = c^2 \iiint \nabla^2\phi\,dx\,dy\,dz = c^2 \iint \frac{d\phi}{dr}\,dS,$$

or, writing $dS = r^2 d\varpi$, so that $d\varpi$ denotes an elementary solid angle,

$$\frac{d^2}{dt^2} \iiint \phi\,dx\,dy\,dz = c^2 r^2 \frac{d}{dr} \iint \phi\,d\varpi.$$

Let us differentiate both sides of this equation with respect to r. The left-hand side gives

$$\frac{d^2}{dt^2} \iint \phi\,dS,$$

so that if we write

$$\bar{\phi} = \frac{1}{4\pi} \iint \phi\,d\varpi = \text{mean value of } \phi \text{ over the sphere,}$$

we have

$$r^2 \frac{d^2 \bar{\phi}}{dt^2} = c^2 \frac{d}{dr}\left(r^2 \frac{d\bar{\phi}}{dr}\right),$$

or

$$\frac{d^2 . r\bar{\phi}}{dt^2} = c^2 \frac{d^2 . r\bar{\phi}}{dr^2} \ldots\ldots\ldots\ldots\ldots\ldots(24^*),$$

the solution of which is

$$r\bar{\phi} = F(r - ct) + f(r + ct).$$

The mean value of ϕ over a sphere having any point P of the medium as centre is therefore propagated according to the same laws as a symmetrical spherical disturbance of the air. We see at once that the value of ϕ at P at the time t depends on the mean initial values of ϕ and $\frac{d\phi}{dt}$ over a sphere of radius ct described about P as centre, so that the disturbance is propagated in all directions with uniform velocity c. If the disturbance be confined originally to a finite portion Σ of space, the disturbance at any point P external to Σ will begin after a time $\frac{r_1}{c}$, will last for a time $\frac{r_2 - r_1}{c}$, and will then cease altogether; r_1, r_2 denoting the radii of the spheres described with P as centre, the one just excluding, the other just including Σ.

To express the solution of (23), already virtually obtained, in an analytical form, let the values of ϕ and $\frac{d\phi}{dt}$, when $t = 0$, be

---

* This result was obtained, in a different manner, by Poisson, *J. de l'Ecole Polytechnique*, 14me cahier (1807), pp. 334—338. The remark that it leads at once to the complete solution of (23), first given by Poisson, *Mém. de l'Acad. des Sciences*, t. 3 (1818—19), is due to Liouville, *J. de Math.* 1856, pp. 1—6. The above references are taken from Liouville's paper.

The equation (24) may be proved also as follows. Suppose an infinite number of systems of rectangular axes arranged uniformly about any point P of the fluid as origin, and let ϕ_1, ϕ_2, ϕ_3, &c. be the velocity-potentials of motions which are the same with respect to these systems as the original motion ϕ is with respect to the system x, y, z. If then $\bar{\phi}$ denote the mean value of the functions ϕ_1, ϕ_2, ϕ_3, &c., $\bar{\phi}$ will be the velocity-potential of a motion symmetrical with respect to the point P, and will therefore satisfy (12). The value of $\bar{\phi}$ at a distance r from P will evidently be the same thing as the mean value of ϕ over a sphere of radius r described about P as centre.

$$\phi = \psi(x, y, z), \quad \frac{d\phi}{dt} = \chi(x, y, z) \ldots\ldots\ldots(25).$$

The mean initial values of these quantities over a sphere of radius r described about (x, y, z) as centre are

$$\bar{\phi} = \frac{1}{4\pi} \iint \psi(x + lr, \; y + mr, \; z + nr) \, d\varpi,$$

$$\frac{d\bar{\phi}}{dt} = \frac{1}{4\pi} \iint \chi(x + lr, \; y + mr, \; z + nr) \, d\varpi,$$

where l, m, n denote the direction-cosines of any radius of this sphere, and $d\varpi$ the corresponding elementary solid angle. Comparing with Art. 168, we see that the value of ϕ at any subsequent time t is

$$\phi = \frac{1}{4\pi} \frac{d}{dt} \cdot t \iint \psi(x + lct, \; y + mct, \; z + nct) \, d\varpi$$

$$+ \frac{t}{4\pi} \iint \chi(x + lct, \; y + mct, \; z + nct) \, d\varpi \ldots\ldots(26).$$

171. We have so far assumed the velocity and the condensation to be so small that their squares and products may be neglected. The results obtained on this supposition are indeed sufficiently accurate for most purposes; but it is worth while to notice briefly the solutions of the exact equations of motion of plane waves which have been obtained, independently and by different methods by Earnshaw[*] and Riemann[†].

Riemann's Method.

Riemann starts from the ordinary Eulerian form of the equations of motion and continuity, which may be written

$$\frac{du}{dt} + u \frac{du}{dx} = -\frac{dp}{d\rho} \frac{d \log \rho}{dx} \ldots\ldots\ldots\ldots(27),$$

$$\frac{d \log \rho}{dt} + u \frac{d \log \rho}{dx} = -\frac{du}{dx} \ldots\ldots\ldots\ldots(28).$$

---

[*] *Phil. Trans.* 1860.
[†] *Gött. Abh.* c. 8, 1860. Reprinted in *Werke*, p. 145.

Multiplying (28) by $\pm \sqrt{\dfrac{dp}{d\rho}}$, adding to (27), and writing for shortness

$$\int \sqrt{\frac{dp}{d\rho}}\, d \log \rho = f(\rho),$$

$$\left.\begin{array}{l} f(\rho) + u = 2r \\ f(\rho) - u = 2s \end{array}\right\} \dots\dots\dots\dots\dots(29),$$

we obtain

$$\frac{dr}{dt} = -\left(u + \sqrt{\frac{dp}{d\rho}}\right)\frac{dr}{dx},$$

$$\frac{ds}{dt} = -\left(u - \sqrt{\frac{dp}{d\rho}}\right)\frac{ds}{dx},$$

whence

$$dr = \frac{dr}{dx}\left\{dx - \left(u + \sqrt{\frac{dp}{d\rho}}\right)dt\right\} \dots\dots\dots(30),$$

$$ds = \frac{ds}{dx}\left\{dx - \left(u - \sqrt{\frac{dp}{d\rho}}\right)dt\right\} \dots\dots\dots(31).$$

If the values of r and s can be found, those of u and ρ follow at once from (29). Now (30) and (31) shew that r is constant for a geometrical point moving with velocity

$$\frac{dx}{dt} = u + \sqrt{\frac{dp}{d\rho}},$$

whilst s is constant for a point moving with velocity

$$\frac{dx}{dt} = u - \sqrt{\frac{dp}{d\rho}}.$$

Hence any given value of r travels forward with the velocity $\sqrt{\dfrac{dp}{d\rho}} + u$, and any given value of s backwards with the velocity $\sqrt{\dfrac{dp}{d\rho}} - u$.

These results enable us, if not to calculate, still to understand the character of the motion in any given case. Thus let us suppose that the initial disturbance is confined to the space between the values a and b of x; so that we have initially for

$$x < a, \quad r = r_1, \quad s = s_1,$$

and for

$$x > b, \quad r = r_2, \quad s = s_2,$$

L.　　　　　　　　　　　　　　　　　　　14

where r_1, r_2, s_1, s_2 are constant. The region within which r is variable advances, that within which s is variable recedes, so that after a time these regions separate, and leave between them a space in which $r = r_1$, $s = s_2$, and which is therefore free from disturbance. The original disturbance is thus split up into two waves travelling in opposite directions. In the advancing wave $s = s_2$, and therefore $u = f(\rho) - 2s_2$, so that both density and particle-velocity advance at the rate $\sqrt{\dfrac{dp}{d\rho}} + u$. This velocity of propagation is greater, the greater the value of ρ. The law of progress of the wave may be illustrated by drawing a curve with x as abscissa and ρ as ordinate, and making every point of this curve move forward with the above velocity. It appears that those parts move fastest for which the ordinates are greatest, so that finally points with larger overtake points with smaller ordinates, and the curve becomes at some point perpendicular to x. The functions $\dfrac{d\rho}{dx}$, $\dfrac{du}{dx}$ are then infinite, and the above formulæ are no longer applicable. We have in fact a 'bore,' or wave of discontinuity. Compare Art. 152.

Earnshaw's Method.

172. The same results follow from Earnshaw's investigation; which is however somewhat less general in that it embraces waves travelling in one direction only. If for simplicity we suppose p and ρ to be connected by Boyle's law

$$p = c^2 \rho,$$

the equation (4) is replaced by

$$\frac{d^2 \xi}{dt^2} = c^2 \frac{\dfrac{d^2 \xi}{dx^2}}{\left(1 + \dfrac{d\xi}{dx}\right)^2},$$

or, writing $y = x + \xi$, so that y denotes the absolute position at time t of the particle x,

$$\frac{d^2 y}{dt^2} = c^2 \frac{d^2 y}{dx^2} \div \left(\frac{dy}{dx}\right)^2 \quad \ldots\ldots\ldots\ldots\ldots(32).$$

This is satisfied by $\dfrac{dy}{dt} = f\left(\dfrac{dy}{dx}\right)$,

provided $\left\{f'\left(\dfrac{dy}{dx}\right)\right\}^2 = c^2 \div \left(\dfrac{dy}{dx}\right)^2$;

so that a first integral of (32) is

$$\frac{dy}{dt} = C \pm c \log \frac{dy}{dx} \quad\dots\dots\dots\dots\dots (33).$$

To obtain the complete solution of (32) we must* eliminate a between the equations

$$\left. \begin{array}{l} y = ax + (C \pm c \log a)\, t + \phi(a) \\ 0 = ax \pm ct + a\phi'(a) \end{array} \right\} \quad\dots\dots\dots\dots (34).$$

Now $\dfrac{dy}{dx} = \dfrac{\rho_0}{\rho}$,

so that if u be the velocity of the particle x, we have

$$u = \frac{dy}{dt} = C \pm c \log \frac{\rho_0}{\rho},$$

whence $\rho = \rho_0 e^{\mp \frac{u-C}{c}}.$

In the parts of the fluid not yet reached by the wave we have $\rho = \rho_0$, $u = 0$. Hence we have $C = 0$, and therefore

$$\rho = \rho_0 e^{\mp \frac{u}{c}} \dots\dots\dots\dots\dots\dots (35),$$

$$y = ax \pm c \log a \cdot t + \phi(a) \quad\dots\dots\dots\dots (36),$$

$$0 = ax \pm ct + a\phi'(a) \quad\dots\dots\dots\dots (37).$$

The function ϕ is determined from the initial circumstances by the equation

$$x = -\phi'\left(\frac{\rho_0}{\rho}\right).$$

To obtain results independent of the form of the wave let us take two particles, which we distinguish by suffixes, so related that the value of ρ which obtains for the first particle at the time t_1 is found at the second particle at the time t_2.

* Boole, *Differential Equations*, c. 14.

The value of $\alpha \left(= \dfrac{\rho_0}{\rho}\right)$ is the same for both these particles, so that we have, by (36) and (37),

$$y_2 - y_1 = \alpha(x_2 - x_1) \pm c \log \alpha \,(t_2 - t_1),$$
$$0 = \alpha\,(x_2 - x_1) \pm c\,(t_2 - t_1).$$

The latter equation may be written

$$\frac{x_2 - x_1}{t_2 - t_1} = \mp c\,\frac{\rho}{\rho_0} \quad\dots\dots\dots\dots\dots\dots(38),$$

which shews that any value ρ of the density is propagated from particle to particle with the velocity $c\dfrac{\rho}{\rho_0}$. The rate of propagation in space is given by $\dfrac{y_2 - y_1}{t_2 - t_1}$, viz. it is $\pm c \log \alpha \mp \dfrac{c}{\alpha}$, or

$$\mp c\,\frac{\rho}{\rho_0} + u \quad\dots\dots\dots\dots\dots\dots\dots(39).$$

For a wave travelling in the positive direction we must take the lower signs throughout. If it be one of condensation (*i.e.* $\rho > \rho_0$), u is, by (35), positive. We see as before that the denser parts of the wave gain continually on the rarer, and at length overtake them, when a bore is formed, and the subsequent motion is beyond the scope of this analysis.

Eliminating x between (36) and (37), and writing for $c \log \alpha$ its value $- u$, we find for a wave travelling in the positive direction,

$$y = (c + u)t + F(x),$$

which may, in virtue of (35), be written

$$u = f\{y - (c + u)t\}.$$

This formula is due to Poisson[*]. Its interpretation, leading of course to the same results as before, was discussed and illustrated by Stokes and Airy, in the *Philosophical Magazine*, in 1848-9.

The theoretical result that violent sounds are propagated faster

---

[*] *Journ. de l'Ecole Polyt.* t. 12, cahier 14, p. 319. Quoted by Earnshaw and Rankine.

than gentle ones is confirmed by the remarkable experiments of Regnault* on the motion of sound in pipes.

173.　The above investigations shew that a plane wave of air experiences in general a gradual change of type as it proceeds. It is worth while to inquire under what circumstances a wave could be propagated without change.　Let A, B be two points of an ideal tube of unit section drawn in the direction of propagation, and let the values of the pressure, the density, and the particle-velocity at A and B be denoted by p_1, ρ_1, u_1, and p_2, ρ_2, u_2, respectively.　If as in Art. 152 we impress on everything a velocity c equal and opposite to that of the waves we reduce the problem to one of steady motion.　Since the same amount of matter now crosses in unit time each section of the tube, we have

$$\rho_1(c - u_1) = \rho_2(c - u_2) = \text{const.} = m, \text{ say}\ldots\ldots\ldots(40).$$

The quantity m, denoting the mass swept past in unit time by a plane moving with the wave, in the original motion, is called by Rankine† the 'mass-velocity' or 'somatic velocity' of the wave.

Again, the total force acting on the mass included between A and B is $p_2 - p_1$, and the gain per unit time of momentum of this mass is

$$m(c - u_1) - m(c - u_2).$$

Hence　　　　　　　$$p_2 - p_1 = m(u_2 - u_1)\ldots\ldots\ldots\ldots\ldots(41).$$

Combined with (40) this gives

$$p_1 + m^2 s_1 = p_2 + m^2 s_2\ldots\ldots\ldots\ldots\ldots(42),$$

where, still following Rankine, we denote by $s\ [= \rho^{-1}]$ the 'bulkiness,' *i.e.* the volume per unit mass, of the substance.　Hence the variations in pressure and bulk experienced by any small portion of the medium as the wave passes over it must be such that

$$p + m^2 s = \text{const}\ldots\ldots\ldots\ldots\ldots(43).$$

This condition is not fulfilled by any known substance, whether at constant temperature, or when free from gain or loss of heat by

---

* *Mém. de l'Acad.* t. 37.　Quoted in Wüllner's *Experimental physik*, t. 1, p. 685.　An abstract of the experiments is printed at the end of the second edition of Tyndall's *Sound*.

† *Phil. Trans.* 1870.

conduction and radiation. " In order then that permanency of type may be possible in a wave of longitudinal disturbance, there must be both change of temperature and conduction of heat during the disturbance."

Rankine, in the paper referred to, considers it unlikely that the conduction of heat ever takes place in such a way as accurately to maintain the relation (43), except in the case of a wave of sudden disturbance, where we have adjacent portions of the medium at a finite interval of temperature.

This latter kind of wave is of interest because, as we have seen, any disturbance however started tends ultimately to become discontinuous. The simplest case is that in which there is no variation in the properties of the medium except at the plane of discontinuity. If p, s, u denote the values of the pressure, the bulkiness, and the particle-velocity behind, P, S, U those in front of the discontinuity, the conditions to be satisfied are obtained by making p_1, s_1, $u_1 = p$, s, u, and p_2, s_2, $u_2 = P$, S, U, respectively, in the above formulæ. We find

$$m = \sqrt{\frac{p - P}{S - s}},$$

and if further $U = 0$ so that the medium is at rest in front of the wave,

$$c = mS = S\sqrt{\frac{p - P}{S - s}},$$

and $$u = m(S - s) = \pm \sqrt{(p - P)(S - s)},$$

the upper or lower sign being taken according as S is greater or less than s, i.e. according as the wave is one of sudden compression or of sudden rarefaction.

The mathematical possibility of, and the conditions for, a wave of discontinuity were first pointed out by Stokes.

CHAPTER IX.

VISCOSITY.

174. WE now proceed to take account of the resistance to distortion, known as 'viscosity' or 'internal friction,' which (Art. 2) is exhibited by all actual fluids in motion, but which we have hitherto neglected. The methods we shall employ are of necessity the same as are applicable to the resistance to distortion, known as 'elasticity,' which is experienced in the case of solid bodies. The two classes of phenomena are of course physically distinct, the latter depending on the actual changes of shape produced, the former on the rate of change of shape, but the mathematical methods appropriate to them are to a great extent identical.

175. Let p_{xx}, p_{xy}, p_{xz} denote the components parallel to x, y, z, respectively, of the force per unit area exerted at the point $P(x, y, z)$ across a plane perpendicular to x, between the two portions of fluid on opposite sides of it; and let p_{yx}, p_{yy}, p_{yz}, and p_{zx}, p_{zy}, p_{zz} have similar meanings with respect to planes perpendicular to y and z respectively*. It is shewn in the theory of Elastic Solids that these nine quantities, which completely specify the state of stress at the point P, are not all independent, but that

$$p_{yz} = p_{zy}, \quad p_{zx} = p_{xz}, \quad p_{xy} = p_{yx} \dots\dots\dots\dots(1).$$

We need not here reproduce the proof of these relations, as their truth will appear from the expressions for p_{yz}, &c. to be given below.

---

* As is usual in the theory of Elasticity we reckon a *tension* positive, a *pressure* negative. Thus in the case of a fluid in equilibrium we have

$$p_{xx} = p_{yy} = p_{zz} = -p.$$

To form the equations of motion let us take, as in Art. 6, an element $dx\,dy\,dz$ having its centre at P, and resolve the forces acting on it parallel to x. Taking first the pair of faces perpendicular to x, the difference of the tensions on these parallel to x will be $\dfrac{dp_{xx}}{dx}\,dx\,.\,dy\,dz$. From the other two pairs we obtain $\dfrac{dp_{yx}}{dy}\,dy\,.\,dz\,dx$ and $\dfrac{dp_{zx}}{dz}\,dz\,.\,dx\,dy$, respectively. Hence, with our usual notation,

and, similarly,

$$\left.\begin{aligned}
\rho\,\frac{\partial u}{\partial t} &= \rho X + \frac{dp_{xx}}{dx} + \frac{dp_{yx}}{dy} + \frac{dp_{zx}}{dz}\;;\\[4pt]
\rho\,\frac{\partial v}{\partial t} &= \rho Y + \frac{dp_{xy}}{dx} + \frac{dp_{yy}}{dy} + \frac{dp_{zy}}{dz}\;,\\[4pt]
\rho\,\frac{\partial w}{\partial t} &= \rho Z + \frac{dp_{xz}}{dx} + \frac{dp_{yz}}{dy} + \frac{dp_{zz}}{dz}\;.
\end{aligned}\right\}\quad\ldots\ldots\ldots\ldots(2).$$

176. It appears from Arts. 2, 3 that the deviation of the state of stress denoted by p_{xx}, p_{xy}, &c. from one of uniform pressure in all directions depends entirely on the motion of distortion of the fluid in the neighbourhood of P, i.e. on the quantities a, b, c, f, g, h by which this distortion was in Art. 38 shewn to be specified. Before endeavouring to express p_{xx}, p_{xy}, &c. as functions of these quantities, it is convenient to establish certain formulæ of transformation.

Let us draw Px', Py', Pz' in the directions of the principal axes of distortion at P, and let a', b', c' be the rates of extension along these lines. Further let the mutual configuration of the two sets of axes, x, y, z and x', y', z', be specified in the usual manner by the annexed scheme of direction-cosines.

	x	y	z
x'	$l_1,$	$m_1,$	$n_1,$
y'	$l_2,$	$m_2,$	$n_2,$
z'	$l_3,$	$m_3,$	$n_3.$

We have, then,

$$\frac{du}{dx} = \left(l_1\frac{d}{dx'} + l_2\frac{d}{dy'} + l_3\frac{d}{dz'} \right)(l_1 u' + l_2 v' + l_3 w')$$

$$= l_1^2\frac{du'}{dx'} + l_2^2\frac{dv'}{dy'} + l_3^2\frac{dw'}{dz'},$$

or

Similarly

$$\left.\begin{aligned}a &= l_1^2 a' + l_2^2 b' + l_3^2 c'. \\ b &= m_1^2 a' + m_2^2 b' + m_3^2 c', \\ c &= n_1^2 a' + n_2^2 b' + n_3^2 c'.\end{aligned}\right\} \quad \ldots\ldots\ldots\ldots (3).$$

We notice that

$$a + b + c = a' + b' + c' \ldots\ldots\ldots\ldots\ldots (4),$$

an invariant, as it should be. See Art. 8. Again

$$\frac{dw}{dy} + \frac{dv}{dz} = \left(m_1 \frac{d}{dx'} + m_2 \frac{d}{dy'} + m_3 \frac{d}{dz'}\right)(n_1 u' + n_2 v' + n_3 w')$$

$$+ \left(n_1 \frac{d}{dx'} + n_2 \frac{d}{dy'} + n_3 \frac{d}{dz'}\right)(m_1 u' + m_2 v' + m_3 w'),$$

or

Similarly

$$\left.\begin{aligned}f &= m_1 n_1 a' + m_2 n_2 b' + m_3 n_3 c'. \\ g &= n_1 l_1 a' + n_2 l_2 b' + n_3 l_3 c', \\ h &= l_1 m_1 a' + l_2 m_2 b' + l_3 m_3 c'.\end{aligned}\right\} \quad \ldots\ldots\ldots\ldots (5).$$

177. From the symmetry of the circumstances it is plain that the stresses exerted at P across the planes $y'z'$, $z'x'$, $x'y'$ are wholly perpendicular to these planes. Let us denote them by p_1, p_2, p_3, respectively. In the figure of Art. 3 let ABC be a plane drawn perpendicular to x, infinitely close to P, meeting the axes of x', y', z' in A, B, C, respectively; and let Δ denote the area ABC. The areas of the remaining faces of the tetrahedron $PABC$ will then be $l_1\Delta$, $l_2\Delta$, $l_3\Delta$. Resolving parallel to x the forces acting on the tetrahedron we find

$$p_{xx}\Delta = p_1 l_1 \Delta \cdot l_1 + p_2 l_2 \Delta \cdot l_2 + p_3 l_3 \Delta \cdot l_3;$$

the external impressed forces and the resistances to acceleration being omitted for the same reason as in Art. 3. Hence

from which we write down, by symmetry,

$$\left.\begin{aligned}p_{xx} &= p_1 l_1^2 + p_2 l_2^2 + p_3 l_3^2, \\ p_{yy} &= p_1 m_1^2 + p_2 m_2^2 + p_3 m_3^2, \\ p_{zz} &= p_1 n_1^2 + p_2 n_2^2 + p_3 n_3^2.\end{aligned}\right\} \quad \ldots\ldots\ldots\ldots (6).$$

We notice that

$$p_{xx} + p_{yy} + p_{zz} = p_1 + p_2 + p_3 = -3p, \text{ say} \ldots\ldots\ldots (7),$$

so that p denotes the average pressure about the point P.

Again, resolving parallel to y we find

$$p_{xy} = p_1 l_1 m_1 + p_2 l_2 m_2 + p_3 l_3 m_3,$$

whence we write down, by symmetry,

$$\left.\begin{array}{l} p_{yz} = p_1 m_1 n_1 + p_2 m_2 n_2 + p_3 m_3 n_3, \\ p_{zx} = p_1 n_1 l_1 + p_2 n_2 l_2 + p_3 n_3 l_3. \end{array}\right\} \quad \cdots\cdots\cdots\cdots(8).$$

The truth of the relations (1) follows from (8) by symmetry. The student should notice the analogy of (3) and (5) with (6) and (8) respectively.

If in the same figure (Art. 3) we suppose PA, PB, PC to be drawn parallel to x, y, z, respectively, and ABC to be any plane near P whose direction-cosines are l, m, n, we find, in exactly the same manner, for the components of the stress exerted across this plane, the values

$$lp_{xx} + mp_{xy} + np_{xz}, \quad lp_{yx} + mp_{yy} + np_{yz}, \quad lp_{zx} + mp_{zy} + np_{zz} \cdots (9),$$

respectively.

178.   Now p_1, p_2, p_3 differ from $-p$ by quantities depending on the motion of distortion, which are therefore functions of a', b', c' only; and if a', b', c' be small we may suppose these functions to be linear.   We write therefore

$$\left.\begin{array}{l} p_1 = -p + \lambda(a' + b' + c') + 2\mu a', \\ p_2 = -p + \lambda(a' + b' + c') + 2\mu b', \\ p_3 = -p + \lambda(a' + b' + c') + 2\mu c', \end{array}\right\} \quad \cdots\cdots\cdots\cdots(10),$$

where λ, μ are constants, this being plainly the most general assumption consistent with the above suppositions, and with symmetry.   Substituting these values of p_1, p_2, p_3 in (6) and (8), and making use of (3) and (5), we find

$$p_{xx} = -p + \lambda(a + b + c) + 2\mu a, \quad \&\text{c.}, \quad \&\text{c.},$$

$$p_{yz} = 2\mu f, \quad \&\text{c.}, \quad \&\text{c.}$$

But from the definition (7) of p we must have

$$3\lambda + 2\mu = 0,$$

whence, finally, introducing the values of a, b, c, &c. from Art. 38, we have

$$p_{xx} = -p - \tfrac{2}{3}\mu \left(\frac{du}{dx} + \frac{dv}{dy} + \frac{dw}{dz}\right) + 2\mu \frac{du}{dx}, \ \&c., \&c.,$$

$$p_{\mu} = \mu \left(\frac{dw}{dy} + \frac{dv}{dz}\right) = p_{zy}, \ \&c., \&c. \qquad \Bigg\rbrace \quad \ldots \ldots (11).$$

The quantity μ is called the 'coefficient of viscosity.' Its physical meaning is as follows. If we conceive the fluid to be moving in a series of horizontal planes, the velocity being in direction everywhere the same, and in magnitude proportional to the distance from some fixed horizontal plane, each stratum of fluid will then exert on the one above it a tangential retarding force whose amount per unit area is μ times the upward rate of variation of the velocity.

If $[M]$, $[L]$, $[T]$ denote the units of mass, length, and time, the dimensions of the p's are $[ML^{-1}T^{-2}]$, and those of a, b, c, &c. are $[T^{-1}]$, so that the dimensions of μ are $[ML^{-1}T^{-1}]$.

The effect of viscosity in modifying the motion of a fluid depends however, as will appear from the examples given below, not so much on the value of μ as on that of $\dfrac{\mu}{\rho}$, $= \mu'$, say. The quantity μ' is called by Maxwell the 'kinematic coefficient of viscosity.' Its dimensions are $[L^2 T^{-1}]$.

It is beyond our province to discuss the various experiments which have been instituted with a view to determining the values of μ, or of μ', for different substances. We may state, however, that μ is found in gases to be independent of the density, and to increase with the temperature, its value being, according to Maxwell, proportional to the temperature measured from the zero of the air-thermometer, e.g. Maxwell finds for air

$$\mu = \cdot 0001878 \, (1 + \cdot 00366 \, \theta),$$

the units of length, mass, and time being the centimeter, gramme, and second, and the temperature θ being expressed in the centigrade scale.

The value of μ for water is, according to the experiments of Helmholtz and Piotrowski,

$$\mu = \cdot 014061$$

at a temperature of $24\cdot5°C$, the units being the same as before. The viscosity of liquids diminishes rapidly with the increase of temperature.

179. Let us calculate the rate at which the stresses acting on a rectangular element $dx\,dy\,dz$ having its centre at (x, y, z) do work in changing its shape and volume. The two yz-faces give

$$\left(p_{xx}\frac{du}{dx}+p_{xy}\frac{dv}{dx}+p_{xz}\frac{dw}{dx}\right)dx\,dy\,dz;$$

and combining with this the similar expressions due to the other pairs of faces we obtain

$$(p_{xx}a + p_{yy}b + p_{zz}c + 2p_{yz}f + 2p_{zx}g + 2p_{xy}h)\,dx\,dy\,dz;$$

or, by (11),

$$-p\,(a + b + c)\,dx\,dy\,dz$$
$$+ \{-\tfrac{2}{3}\mu(a+b+c)^2 + 2\mu(a^2+b^2+c^2+2f^2+2g^2+2h^2)\}\,dx\,dy\,dz$$
$$\dots\dots\dots\dots(12^*).$$

The first line in this formula expresses the rate at which a uniform pressure p works in changing the volume of the element at the rate $a + b + c$. Since this term changes sign with a, b, c, it follows that if p be a function of the density, (and therefore of the volume of the element,) the work done by p during a compression is restored during an expansion, and *vice versa*†. The expression in the second line of (12), however, does not change sign with a, b, c, and represents a real loss of energy. The internal friction therefore involves a dissipation of energy at the rate F per unit volume, where

$$F = -\tfrac{2}{3}\mu(a+b+c)^2 + 2\mu(a^2 + b^2 + c^2 + 2f^2 + 2g^2 + 2h^2)\dots(13).$$

F may be called, as it is by Lord Rayleigh‡, the 'dissipation-function.' We notice that

$$p_{xx} = -p + \tfrac{1}{2}\frac{dF}{da}, \quad \&c., \quad \&c.,$$

$$p_{yz} = \tfrac{1}{4}\frac{dF}{df}, \quad \&c., \quad \&c.$$

---

* Stokes, *Camb. Trans.* Vol. IX. p. 58.

† The assumption that the intrinsic energy of an element of (perfect) fluid depends only on its volume is the basis of Lagrange's treatment of Hydrodynamics in the *Mécanique Analytique*.

‡ *Proc. Lond. Math. Soc.* June 12, 1873.

180.   If we substitute from (11) in the equations of motion (2), we find that the latter may be written

$$\rho \frac{\partial u}{\partial t} = \rho X - \frac{dp}{dx} + \tfrac{1}{3}\mu \frac{d}{dx}\left(\frac{du}{dx} + \frac{dv}{dy} + \frac{dw}{dz}\right) + \mu\left(\frac{d^2u}{dx^2} + \frac{d^2u}{dy^2} + \frac{d^2u}{dz^2}\right) \Bigg\}\ (14).$$

$$\text{\&c., \&c.}$$

When the fluid is incompressible these become

$$\left.\begin{aligned}
\rho \frac{\partial u}{\partial t} &= \rho X - \frac{dp}{dx} + \mu\nabla^2 u, \\
\rho \frac{\partial v}{\partial t} &= \rho Y - \frac{dp}{dy} + \mu\nabla^2 v, \\
\rho \frac{\partial w}{\partial t} &= \rho Z - \frac{dp}{dz} + \mu\nabla^2 w,
\end{aligned}\right\} \dots\dots\dots\dots(15),$$

where ∇^2 has its usual meaning.

In the case of compressible fluids (gases) we may by writing

$$p' = p - \tfrac{1}{3}\mu\left(\frac{du}{dx} + \frac{dv}{dy} + \frac{dw}{dz}\right)$$

reduce (14) to the form (15), but since in all probability the laws of Boyle and Charles hold with regard to p but not to p', there is no real gain of simplicity.

The equations (14) or (15) have been obtained by Navier, Cauchy, Poisson*, and others, on various considerations as to the nature and mutual action of the ultimate molecules of fluids.   The method adopted above, which seems due in principle to de Saint-Venant and Stokes†, is independent of all hypotheses of this kind, but it must be remembered that it involves the assumption that $p_{xx} + p$, p_{xy}, &c. are *linear* functions of the coefficients of distortion. Hence although (14) and (15) may apply with great accuracy to cases of slow motion‡, we have no assurance of their validity in other cases.

It may be remarked, however, that the calculations of Maxwell§, who has investigated the viscosity of gases on the assump-

* For references see Stokes, *B. A. Reports*, 1846, and O. E. Meyer, *Crelle*, t. 59.
† *Camb. Trans.* Vol. viii. 1845.
‡ That they do so is in fact shewn by the experiments of Maxwell and of Helmholtz and Piotrowski.
§ *Phil. Trans.* 1867.

tions of the kinetic theory, point to the validity of the formulæ (11) within very wide limits as to the values of a, b, c, &c. Since the postulates of this theory are in the main highly probable, we are warranted in regarding the equations (14) as sensibly accurate for all ordinary cases of motion of gases.

181. We have still to consider the conditions to be satisfied at a boundary; at the common surface, for instance, of the fluid and of a solid with which it is in contact. It is found that in many cases* the particles of fluid in contact with the solid move with the latter, so that if u, v, w and u', v', w' denote the component velocities of contiguous elements of the fluid and the solid respectively, we have

$$u = u', \qquad v = v', \qquad w = w' \dots \dots \dots \dots \dots (16).$$

In other cases it appears that there is a finite slipping of the fluid past the surface of the solid. The boundary-condition may then be obtained as follows. Considering the motion of a small film of fluid, of thickness infinitely small compared with its lateral dimensions, in contact with the solid, we see that the tangential stress on its inner surface must ultimately balance the force exerted on its outer surface by the solid. The former stress may be calculated from (9); the latter may be taken as directly opposing the velocity of the fluid relatively to the solid, and as approximately proportional to this relative velocity, *provided it be small.* The constant (β, say,) which expresses the ratio of the tangential force to the relative velocity may be called the 'coefficient of sliding friction.'

The conditions to be satisfied at the common surface of two different fluids, or of two portions of the same fluid separated by a surface of discontinuity, may be obtained in the same way.

The boundary-conditions here given are dynamical; the purely kinematical condition of Art. 10 of course obtains here as always.

---

* Stokes (*Camb. Trans.* Vol. IX.) and Maxwell (*Phil. Trans.* 1866) find this to be true of air in contact with slowly vibrating bodies. Helmholtz and Piotrowski infer from their experiments (*Wiener Sitz. B.* t. 40, 1860) that there is little or no slipping for the cases of ether and alcohol in contact with a polished and gilt metal surface. On the other hand, they found that in the case of water, and of most other liquids which they experimented upon, an appreciable amount of slipping took place.

182. We proceed to apply our equations to the treatment of a few particular problems.

Example 1. We take first the flow of a liquid in a straight pipe of uniform circular section, neglecting external impressed forces. If we take the axis of the pipe as the axis of x, and assume $v = 0$, $w = 0$, the second and third equations of (15) give $\frac{dp}{dy} = 0$, $\frac{dp}{dz} = 0$, and the equation of continuity becomes $\frac{du}{dx} = 0$. If we further assume that u is a function of $r (= (y^2 + z^2)^{\frac{1}{2}})$ and t only, the first equation of (15) becomes on transformation of co-ordinates

$$\frac{du}{dt} = -\frac{1}{\rho}\frac{dp}{dx} + \mu' \left(\frac{d^2u}{dr^2} + \frac{1}{r}\frac{du}{dr} \right) \quad\ldots\ldots\ldots\ldots(17).$$

If a be the radius of the pipe, the surface-condition is, Art. 181,

$$-\mu\frac{du}{dr} = \beta u \ldots\ldots\ldots\ldots\ldots\ldots(18),$$

when $r = a$.

In steady motion $\frac{du}{dt} = 0$, so that (17) becomes

$$\frac{d^2u}{dr^2} + \frac{1}{r}\frac{du}{dr} = \frac{1}{\mu'\rho}\frac{dp}{dx}.$$

The left-hand side of this equation does not involve x, and the right-hand side does not involve r, so that each must be constant, $= A$, say. This gives

$$\frac{d^2u}{dr^2} + \frac{1}{r}\frac{du}{dr} = A \ldots\ldots\ldots\ldots\ldots(19),$$

$$\frac{dp}{dx} = \mu'\rho A \ldots\ldots\ldots\ldots\ldots\ldots(20).$$

The solution of (19) is

$$u = \tfrac{1}{4} A r^2 + B \log r + C \ldots\ldots\ldots\ldots(21),$$

where B, C are arbitrary constants. Since the velocity at the centre of the pipe is necessarily finite, we must have $B = 0$; C is then determined by (18), so that

$$u = \tfrac{1}{4} A (r^2 - a^2) - \tfrac{1}{2}\frac{\mu a}{\beta} A.$$

The total flux across any section of the pipe is

$$\int_0^a u \, 2\pi r \, dr = -\tfrac{1}{8}\pi a^4 A - \tfrac{1}{2}\frac{\pi\mu a^3}{\beta} A.$$

If now in a length l of the pipe the pressure fall from p_1 to p_2, we have, by (20),

$$p_1 - p_2 = -\mu'\rho A l,$$

and the flux is

$$\tfrac{1}{8}\frac{\pi a^4}{\mu'\rho}\frac{p_1 - p_2}{l} + \tfrac{1}{2}\frac{\pi a^3}{\beta}\frac{p_1 - p_2}{l} \dots\dots\dots\dots(22).$$

If $\beta = \infty$, there is no slipping at the boundary, and (22) then reduces to its first term

$$\tfrac{1}{8}\frac{\pi a^4}{\mu'\rho}\frac{p_1 - p_2}{l} \dots\dots\dots\dots\dots(22a).$$

Poiseuille found in his experiments* on the flow of water through capillary tubes that the time of efflux of a given quantity of water was directly as the length of the tube, inversely as the difference of pressure at the two ends, and inversely as the fourth power of the diameter. These results agree with (22a).

A comparison of the formula (22a) with experiments of this kind would give the means of determining μ.

183. *Example* 2. To investigate the effect of internal friction on the motion of plane waves of sound. If as in Art. 166 we neglect the squares and products of small quantities, the equation of motion is

$$\frac{du}{dt} = -\frac{1}{\rho}\frac{dp}{dx} + \tfrac{4}{3}\mu'\frac{d^2u}{dx^2},$$

or, writing $u = \dfrac{d\xi}{dt}$, and eliminating p as before by means of the relation $p \propto \rho_\gamma$,

$$\frac{d^2\xi}{dt^2} = c^2\frac{d^2\xi}{dx^2} + \tfrac{4}{3}\mu'\frac{d^3\xi}{dx^2 dt} \dots\dots\dots\dots(23).$$

To fix the ideas let us suppose that at the plane $x = 0$ a simple vibration of period $\dfrac{1}{n}$ is kept up, so that

$$\xi = A \sin 2\pi nt \dots\dots\dots\dots\dots(24),$$

---

* *Mém. des Sav. Etrangers*, t. 9, 1846. Quoted by Wüllner, *Experimental physik*, t. 1, p. 329.

when $x = 0$. Now (23) is satisfied by the sum of any number of terms of the form

$$\xi = Ce^{\alpha x + \beta t} \dots\dots\dots\dots\dots(25),$$

provided

$$\alpha^2 = c^2\beta^2 + \tfrac{4}{3}\mu'\alpha\beta^2 \dots\dots\dots\dots(26).$$

To obtain a solution consistent with (24), we write $\alpha = 2\pi ni$, whence

$$-4\pi^2 n^2 = \beta^2 c^2 (1 + \tfrac{8}{3}\pi\mu' ni),$$
$$= \beta^2 c^2 \sigma (\cos 2\epsilon + i \sin 2\epsilon).$$

Here

$$\sigma^2 = 1 + \tfrac{64}{9} \pi^2 \mu'^2 n^2,$$

and 2ϵ is the least positive solution of

$$\tan 2\epsilon = \tfrac{8}{3} \pi\mu' n.$$

Hence

$$\beta = \pm \frac{2\pi ni}{c\sqrt{\sigma}} (\cos \epsilon - i \sin \epsilon).$$

Substituting, and putting (25) in a real form, we find

$$\xi = Ae^{\pm\frac{x}{l}} \sin 2\pi n \left(t \pm \frac{x}{c'} \right) \dots\dots\dots\dots(27),$$

where

$$\frac{1}{l} = \frac{2\pi n}{c\sqrt{\sigma}} \sin \epsilon,$$

and

$$c' = c\sqrt{\sigma} \sec \epsilon.$$

In (27) we must take the upper or the lower sign according as we consider the waves propagated in the direction of x negative, or x positive. We see that the velocity of propagation is increased by the friction, the increase being greater for higher notes than for lower ones. The amplitude of the waves diminishes as they proceed, the diminution being the more rapid the higher the note. The change in the velocity of propagation depends ultimately on the square of μ, that of the amplitude on the first power of μ, so that the latter effect is much more important than the former. Stokes[*] however has shewn that in all ordinary cases the diminution of amplitude due to friction is insignificant compared with that due to spherical divergence.

The equation (27) does not constitute the *complete* solution of (23) subject to the condition (24). In fact we may add any number

[*] *Camb. Phil. Trans.* Vol. ix. p. 94.

of terms of the form (25), provided the coefficients be so chosen as to make

$$\xi = 0, \text{ for } x = 0 \dots\dots\dots\dots(28).$$

Thus, writing $\beta = ki$ in (26), and solving with respect to α, we find

$$\alpha = -\tfrac{2}{3}\mu'k^2 \pm kc'i,$$

where

$$c'^2 = c^2\left(1 - \tfrac{4}{9}\mu'^2\frac{k^2}{c^2}\right) \dots\dots\dots\dots(29).$$

Substituting in (25), and making use of (28), we obtain the solution

$$\xi = \Sigma e^{-\frac{2}{3}\mu'k^2t}(P\cos kc't + Q\sin kc't)\sin kx\dots\dots(30),$$

where the summation embraces all values of k. The coefficients P, Q may be determined by Fourier's method so as to make the sum of (27) and (30) satisfy any arbitrary initial conditions. We see that the effect of the initial circumstances gradually disappears, until finally the only sensible part of the disturbance is that due to the forced vibration maintained at $x = 0$.

It appears from (30) that the effect of friction is to diminish the velocity of propagation of a free wave.

184. *Example* 3. A sphere moves with uniform velocity in an incompressible viscous fluid; to find the force which must be applied to the sphere in order to maintain this motion*.

Take the centre of the sphere as origin, the direction of motion as axis of x. If we impress on the fluid and the solid a velocity V equal and opposite to that of the sphere, the problem is reduced to one of steady motion. Further, assuming the motion to be symmetrical about the axis of x, we may write, with the same notation as in Art. 103,

$$u = \frac{1}{\varpi}\frac{d\psi}{d\varpi}, \quad v = -\frac{1}{\varpi}\frac{d\psi}{dx},$$

whence we have, for the angular velocity ω of a fluid element,

$$2\omega = \frac{dv}{dx} - \frac{du}{dy} = -\frac{1}{\varpi}\left(\frac{d^2\psi}{dx^2} + \frac{d^2\psi}{d\varpi^2} - \frac{1}{\varpi}\frac{d\psi}{d\varpi}\right) \dots\dots(31).$$

If we suppose the motion to be so slow that the squares and pro-

---

* Stokes, *Camb. Phil. Trans.* Vol. IX. p. 48.

ducts of u, v, w, $\dfrac{du}{dx}$, &c. may be neglected, we obtain by elimination of p from (15),

$$\nabla^2 \eta = 0, \quad \nabla^2 \zeta = 0 \dots\dots\dots\dots(32).$$

If, as in Art. 140, we write

$$\eta = -\omega \sin \vartheta, \qquad \zeta = \omega \cos \vartheta,$$

we find on transformation of co-ordinates that the equations (32) reduce to the one equation

$$\frac{d^2\omega}{dx^2} + \frac{d^2\omega}{d\varpi^2} + \frac{1}{\varpi}\frac{d\omega}{d\varpi} - \frac{\omega}{\varpi^2} = 0,$$

or

$$\left[\frac{d^2}{dx^2} + \frac{d^2}{d\varpi^2} - \frac{1}{\varpi}\frac{d}{d\varpi}\right]\varpi\omega = 0,$$

whence, substituting the value of ω from (31), we obtain

$$\left[\frac{d^2}{dx^2} + \frac{d^2}{d\varpi^2} - \frac{1}{\varpi}\frac{d}{d\varpi}\right]^2 \psi = 0 \dots\dots\dots(33).$$

If we write

$$x = r\cos\theta, \qquad y = r\sin\theta,$$

this becomes

$$\left[\frac{d^2}{dr^2} + \frac{1}{r^2}\frac{d^2}{d\theta^2} - \frac{1}{r^2}\cot\theta\frac{d}{d\theta}\right]^2 \psi = 0,$$

or

$$\left[\frac{d^2}{dr^2} + \frac{\sin\theta}{r^2}\frac{d}{d\theta}\left(\frac{1}{\sin\theta}\frac{d}{d\theta}\right)\right]^2 \psi = 0 \dots\dots\dots(34).$$

If R, Θ denote the component velocities at any point of the fluid along r, and perpendicular to r in the plane of θ, respectively, we have, from the definition of ψ,

$$R = \frac{1}{r\sin\theta}\frac{d\psi}{r\,d\theta}, \qquad \Theta = -\frac{1}{r\sin\theta}\frac{d\psi}{dr}.$$

The conditions to be satisfied at infinity are

$$R = -V\cos\theta, \qquad \Theta = V\sin\theta \dots\dots\dots(35).$$

If we assume

$$\psi = \sin^2\theta\, f(r),$$

we find that (34) is satisfied provided

$$\left[\frac{d^2}{dr^2} - \frac{2}{r^2}\right]^2 f(r) = 0.$$

To solve this equation we assume

$$f(r) = \Sigma A r^n,$$

which gives $\quad (n+1)(n-2)(n-1)(n-4) = 0,$

so that $\quad f(r) = \dfrac{A}{r} + Br + Cr^2 + Dr^4.$

The conditions (35) are then satisfied by

$$D = 0, \qquad 2C = -V \ldots\ldots\ldots \ldots\ldots\ldots\ldots(36).$$

We have still to introduce the conditions to be satisfied at the surface of the sphere. On the hypothesis of no slipping, these are

$$R = 0, \qquad \Theta = 0 \ldots\ldots\ldots\ldots\ldots\ldots\ldots(37),$$

when $r = a$ (the radius), or

$$f(a) = 0, \qquad f'(a) = 0 \ldots\ldots\ldots\ldots\ldots(38).$$

Combined with (36), these give

$$A = -\tfrac{1}{4}Va^3, \qquad B = \tfrac{3}{4}Va \ldots\ldots\ldots\ldots\ldots(39),$$

but we retain for the present the symbols A, B. The resulting value of ψ is simplified if we restore the problem to its original form by removing the impressed velocity $- V$ in the direction of x, *i.e.* if we add to ψ the term

$$\tfrac{1}{2}Vr^2 \sin^2\theta.$$

Thus $\qquad\qquad \psi = \left(\dfrac{A}{r} + Br\right)\sin^2\theta \ldots\ldots\ldots\ldots\ldots(40),$

whence

$$R = 2\left(\dfrac{A}{r^3} + \dfrac{B}{r}\right)\cos\theta, \quad \Theta = \left(\dfrac{A}{r^3} - \dfrac{B}{r}\right)\sin\theta \ldots\ldots\ldots(41).$$

The resistance experienced by the sphere is most readily calculated by means of the dissipation-function F of Art. 179. Let us take at any point a subsidiary system of rectangular axes, in the directions of R, Θ, and of a normal to the plane of θ, respectively; and let $[a], [b], [c], [f], [g], [h]$ have the same meanings with respect to these axes that a, b, c, f, g, h have with respect to x, y, z. We find by simple calculations

$$[a] = \dfrac{dR}{dr}, \quad [b] = \dfrac{d\Theta}{rd\theta} + \dfrac{R}{r}, \quad [c] = \dfrac{R}{r} + \dfrac{\Theta}{r\tan\theta},$$

$$2[f] = \dfrac{d\Theta}{dr} + \dfrac{dR}{rd\theta} - \dfrac{\Theta}{r}, \qquad [g] = 0, \quad [h] = 0,$$

whence

$$[a] = -2\left(\frac{3A}{r^4} + \frac{B}{r^2}\right)\cos\theta,$$

$$[b] = \left(\frac{3A}{r^4} + \frac{B}{r^2}\right)\cos\theta,$$

$$[c] = \left(\frac{3A}{r^4} + \frac{B}{r^2}\right)\cos\theta,$$

$$2[f] = -\frac{6A}{r^4}\sin\theta.$$

Hence

$$F = 12\mu\left(\frac{3A}{r^4} + \frac{B}{r^2}\right)^2\cos^2\theta + 36\mu\frac{A^2}{r^6}\sin^2\theta.$$

To find the total rate of dissipation of energy we must multiply this by $2\pi r\sin\theta\, r d\theta\, dr$, and integrate with respect to θ between the limits 0 and π, and with respect to r between the limits a and ∞. The final result is

$$16\pi\mu\left(\frac{3A^2}{a^5} + \frac{2AB}{a^3} + \frac{B^2}{a^2}\right) \quad\dots\dots\dots\dots\dots(42).$$

If we substitute from (39), this becomes

$$6\pi\mu a V^2.$$

Now if P be the force which must act on the sphere in order to maintain the motion, the rate at which this force works is PV, whence

$$PV = 6\pi\mu a V^2,$$

or

$$P = 6\pi\mu a V\dots\dots\dots\dots\dots\dots(43).$$

In the case of a sphere of mean density σ falling under the action of gravity, P is the excess of the weight $\frac{4}{3}\pi\sigma a^3 g$ over the buoyancy $\frac{4}{3}\pi\rho a^3 g$, so that the terminal velocity is given by

$$V = \frac{2}{9}\frac{\rho - \sigma}{\mu} g a^2\dots\dots\dots\dots\dots(44).$$

For a globule of water, of ·001 in. radius, falling in air, Stokes finds[*],

$$V = 1·59 \text{ inches per second.}$$

---

[*] The value of μ employed in this calculation was deduced from Baily's experiments on pendulums, and is about half as great as that found by Maxwell (Art. 178). If we accept this latter value, the above value of V must be halved.

For a sphere of one-tenth this size, the value of V would be one hundred times less. The viscosity of the air is therefore amply sufficient to account for the apparent suspension of clouds, &c.

185. If we suppose that there is a finite amount of slipping at the surface of the sphere, the conditions (37) must be modified. Regarding the sphere as in motion, and the fluid at rest at infinity, the conditions to be satisfied when $r = a$ are

$$R = V\cos\theta, \text{ and } 2\mu\,[f] = \beta(\Theta + V\sin\theta),$$

where β is the coefficient of sliding friction. These give

$$\frac{A}{a^3} + \frac{B}{a} = \tfrac{1}{2} V, \quad -6\mu\frac{A}{a^4} = \beta\left(\frac{A}{a^3} - \frac{B}{a} + V\right),$$

whence

$$\frac{A}{a^3} = -\tfrac{1}{4}V \div \left(1 + \frac{3\mu}{\beta a}\right), \quad \frac{B}{a} = \tfrac{3}{4}V\left(1 + \frac{2\mu}{\beta a}\right) \div \left(1 + \frac{3\mu}{\beta a}\right)\dots(45).$$

Substituting these values in (42), we obtain for the resistance experienced by the sphere

$$P = 6\pi\mu a V \frac{1 + 4\dfrac{\mu}{\beta a} + 6\left(\dfrac{\mu}{\beta a}\right)^2}{\left(1 + 3\dfrac{\mu}{\beta a}\right)^2} \dots\dots\dots\dots(46).$$

When $\beta = \infty$ this reduces to (43). Whatever the value of β may be, P always lies between $4\pi\mu a V$ and $6\pi\mu a V$.

NOTES.

NOTE A. ART. 1.

WHEN, as in the present subject, and in the cognate theory of Elastic Solids, we study the changes of shape which a mass of matter undergoes, we begin by considering the whole mass as made up of a large number of very small parts, or 'elements,' and endeavouring to take account of the motion, or the displacement, of each of these.

If we inquire however to what extent this ideal subdivision is to be carried, we are at once brought face to face with questions as to the ultimate structure and properties of matter. We have, in the text, adopted the hypothesis of a continuous structure, in which case the subdivision is without limit.

The truth of this hypothesis is, however, not probable. It is now generally held that all substances are ultimately of a heterogeneous or coarse-grained* structure; that they are in fact built up of discrete bodies or 'molecules,' separated, it may be, by more or less wide intervals. These bodies are far too minute to admit of direct observation, so that we are almost wholly ignorant of their nature and of the manner in which they act on one another. It would therefore be futile to attempt to form the equations of motion of individual molecules, and even if the equations could be formed and integrated the results would not be directly comparable with observation, nor would they even be of interest from the point of view of our present subject. For our object is not to follow the careers of individual molecules, which cannot be traced or identified, but to study the motions of portions of matter which, though very small, are still large enough to be observed, and

* Thomson and Tait, *Natural Philosophy*, Art. 675.

which therefore necessarily consist each of an immense number of molecules.

In order then to establish the fundamental equations in a manner free from special hypothesis as to the structure of matter we adopt the following conventions.

We suppose the 'elements' above spoken of to be such that each of their dimensions is a large multiple of the average distance (d, say,) between the centres of inertia of neighbouring molecules, and also of the average distance (δ, say,) beyond which the direct action of one molecule on another becomes insensible*. The latter proviso is necessary in order that the mutual forces exerted between adjacent elements shall be sensibly proportional to the surfaces across which they act. Observation shews that we may suppose the dimensions of an element to be at the same time so small that the average properties of the constituent molecules vary regularly and continuously as we pass from one element to another. We shall, in what follows, understand by the word 'particle' or 'element' a portion of matter whose dimensions lie within the limits here indicated. The properties of an element surrounding any point P may then be treated as continuous functions of the position of P.

The 'density' at any point of the fluid is now to be defined as the ratio of the mass to the volume of an elementary portion surrounding that point.

The 'velocity at a point' is defined as the velocity of the centre of inertia of an elementary portion taken about that point. This is of course quite distinct from the velocities of the individual molecules, which may, and in all probability do, vary quite irregularly from one molecule to another.

By the 'flux' across an ideal surface situate at any point of the fluid we shall understand the mass of matter which in unit time crosses unit area of the surface, from one side to the other. Matter crossing in the direction opposite to that in which the flux is estimated is here reckoned as negative.

The flux across any surface at any point is equal to the product of the density (ρ) into the velocity (q, say,) estimated in the direction of the normal to the surface. This is sufficiently obvious if the fluid be supposed continuous, or even if it be molecular, provided that in the

---

* It is supposed that in gases d is large compared with δ; the reverse is probably the case in liquids and solids.

latter case we assume the velocities of all the molecules within an element to agree in magnitude and direction. To prove the statement when the velocity is supposed to vary quite irregularly from one molecule to another, we have only to suppose the molecules contained in an elementary space to be grouped according to their velocities, so that the velocities of all the members of any one group shall be sensibly the same in magnitude and direction. Let ρ_1, q_1 be the values of the density, and of the velocity in the direction of the normal to the given surface, corresponding to the first group alone, ρ_2, q_2 the corresponding values for the second group, and so on. The part of the flux due to the first group is $\rho_1 q_1$, that due to the second is $\rho_2 q_2$, and so on, so that the total flux in the given direction is

$$\rho_1 q_1 + \rho_2 q_2 + \ldots\ldots,$$

which is, by the definition of the symbols, $= \rho q$.

Hence the flux across any portion of a surface every point of which moves with the fluid is zero.

The foregoing considerations and definitions apply alike to solids and to fluids. But if we attempt to follow the motion of an element we are met by a difficulty peculiar to our present subject. We cannot assume that the molecules which at any instant constitute an element continue to form a compact group throughout the motion. On the contrary the phenomena of diffusion shew that such an association of molecules is gradually disorganized, some molecules being continually detached from the main body, whilst others find their way into it from without. Thus although the matter included by a small closed surface moving with the fluid is constant in amount, its composition is continually changing. It is true that in liquids this process is exceedingly slow, and might fairly be neglected when regarded from our present point of view. In the case of gases it must however be taken into account, in consequence of the much greater mobility of their molecules.

The phrase 'path of a particle,' often used in the text, must therefore now be understood to mean the path of a geometrical point which moves always with the velocity of the fluid where it happens to be. In the case of a liquid this will represent with considerable accuracy the path of the centre of inertia of a definite portion of matter.

The effect of the foregoing definitions is to replace the original (molecular) fluid by a model, made of an ideal continuous substance, in which only the main features of the motion are preserved. The correspondence of the model to the original is however as yet merely kinematical; thus we cannot assert that the model will, with the same internal forces, work similarly to the original.

Let u, v, w be the component velocities of the fluid, as above defined, at the point (x, y, z), $u + a$, $v + \beta$, $w + \gamma$ those of a particular molecule of mass m in the neighbourhood of that point. Let the symbol Σ denote a summation extending through unit volume; more precisely, let Σ denote the result of a summation through unit volume on the supposition that the properties of the medium are uniform throughout this volume, and the same as at (x, y, z). We have, then, in virtue of the definitions above given

$$\Sigma m = \rho \dots\dots\dots\dots\dots\dots\dots\dots\dots(1),$$

$$\Sigma ma = 0, \quad \Sigma m\beta = 0, \quad \Sigma m\gamma = 0 \dots\dots\dots\dots\dots (2).$$

If we consider a small closed surface moving with the fluid, the symbol $\dfrac{\partial}{\partial t}$, $= \dfrac{d}{dt} + u\dfrac{d}{dx} + v\dfrac{d}{dy} + w\dfrac{d}{dz}$, applied to any function F, expresses the rate of variation of F considered as a property of the included space. On account of the continual exchange of molecules between this space and the surrounding region, this is not necessarily the same thing as the rate of variation of F considered as a property of the matter which happens to occupy this space at the instant in question. Hence we cannot, as in Art. 6, accept

$$\text{mass} \times \frac{\partial u}{\partial t}$$

as a complete expression for the rate of increase of the x-momentum of a moving element. A correction is rendered necessary by the passage to and fro of molecules carrying their momentum with them across the walls of the element. To form the equations of motion it is better, then, to have recourse to the second method, explained in Art. 12, viz. to fix our attention on a particular region of space, and study the changes produced in its properties as well by the flow of matter across its boundaries as by the action of external forces.

Let us take then a rectangular element of space $dx\,dy\,dz$, having its centre at (x, y, z). Let Q denote any property of a molecule which it can carry with it in its motion*. The total rate of increase of the amount of Q in the above space is expressed by

$$\frac{d}{dt} \cdot \Sigma Q\,dx\,dy\,dz \dots\dots\dots\dots\dots\dots\dots\dots(3).$$

---

* The following investigation is substantially that given by Maxwell, *l. c.* Art. 12.

Of this the part due to the flow of matter across the yz-face nearest the origin is

$$\left\{ \Sigma Q\,(u+a) - \tfrac{1}{2}\frac{d}{dx}\,\Sigma Q\,(u+a)dx \right\} dy\,dz,$$

and that due to the opposite face is

$$-\left\{ \Sigma Q\,(u+a) + \tfrac{1}{2}\frac{d}{dx}\,\Sigma Q\,(u+a)\,dx \right\} dy\,dz.$$

Calculating in the same way the parts due to the remaining pairs of faces, we find for the total variation in the amount of Q in the element, due to flow of matter across its walls, the expression

$$-\left\{\frac{d}{dx}\Sigma Q\,(u+a) + \frac{d}{dy}\,\Sigma Q\,(v+\beta) + \frac{d}{dz}\,\Sigma Q\,(w+\gamma)\right\} dx\,dy\,dz \dots (4).$$

In this let us first put $Q = m$, the mass of a molecule. Then equating (3) and (4), and taking account of the relations (1) and (2) we find

$$\frac{d\rho}{dt} + \frac{d\,.\,\rho u}{dx} + \frac{d\,.\,\rho v}{dy} + \frac{d\,.\,\rho w}{dz} = 0,$$

the equation of continuity.

Next let us put $Q = m\,(u+a)$, the x-momentum of a molecule. Change in the amount of momentum contained in the space $dx\,dy\,dz$ is produced not only by the flow of matter across the boundary but also by the action of force from without. The external impressed forces X, Y, Z increase the momentum parallel to x at the rate

$$\Sigma m.X\,dx\,dy\,dz \dots (5);$$

and the effect of the internal forces in the same direction is

$$\left(\frac{dp_{xx}}{dx} + \frac{dp_{yx}}{dy} + \frac{dp_{zx}}{dz}\right) dx\,dy\,dz \dots (6),$$

where we have introduced the general specification of the state of stress at (x, y, z) from Art. 175. Equating then (3) to the sum of (4), (5), and (6), and making use of (1) and (2) as before, we find

$$\frac{d\,.\,\rho u}{dt} + \frac{d\,.\,\rho u^2}{dx} + \frac{d\,.\,\rho uv}{dy} + \frac{d\,.\,\rho uw}{dz}$$

$$= \rho\,.X + \frac{d}{dx}\,(p_{xx} - \Sigma ma^2) + \frac{d}{dy}\,(p_{yx} - \Sigma ma\beta) + \frac{d}{dz}\,(p_{zx} - \Sigma ma\gamma).$$

If we write

$$p'_{xx} = p_{xx} - \Sigma m a^2, \quad \&c., \quad \&c.,$$

$$p'_{yz} = p_{yz} - \Sigma m \beta \gamma, \quad \&c., \quad \&c., \quad .$$

this becomes, in virtue of the equation of continuity,

$$\rho \frac{\partial u}{\partial t} = \rho X + \frac{dp'_{xx}}{dx} + \frac{dp'_{yx}}{dy} + \frac{dp'_{xz}}{dz} \quad \dots\dots\dots\dots(7),$$

which agrees in form with (2) of Art. 175.

The quantities p'_{xx}, p'_{yx}, &c. determine the *apparent* stress of the medium at the point (x, y, z), as distinguished from the actual statical stress specified by p_{xx}, p_{yx}, &c. The apparent stress is what is really observed in all experiments in fluids. It is in fact the stress which must hold at any point of the continuous model above spoken of, in order that the model may work similarly to the original.

In any case where we could assume that the oblique component of the stress across any plane is small enough to be neglected, we should find, calculating the rate of change of the momentum contained in a tetrahedral space such as that in the figure of Art. 3,

$$p'_{xx} = p'_{yy} = p'_{xx}, \; = -p, \; \text{say}.$$

The equations of motion then assume the form (2) of Art. 6.

Lemma. If, in any region, a circuit * or system of circuits A is reconcileable with a system B, and also with a system C, the systems B and C are reconcileable.

For it is possible to connect, in the first place A and B, in the second A and C, by continuous surfaces lying wholly within the region and completely bounded by the lines A, B and A, C respectively. These two surfaces adjoin one another along the lines A, and if these lines be obliterated we obtain a continuous surface bounded by B and C, which are therefore reconcileable.

Suppose now that we have in any region a system of n independent circuits a_1, a_2, ... a_n, such that every other circuit drawn in the region is reconcileable with one or more of these. We shall prove that any other set of n independent circuits b_1, b_2, ... b_n which can be drawn in the region possess the same property.

Let x denote any other circuit drawn in the region. By hypothesis b_1 is reconcileable with one or more of the system a_1, a_2, ... a_n, say with a_1, combined (if necessary) with others of the system; and x is also reconcileable with one or more of the system a_1, a_2, ... a_n. Firstly let us suppose that a_1 is included in the set with which x is reconcileable. We have then a_1 reconcileable with one or more of the system b_1, a_2, a_3, ... a_n, and also with one or more of the system x, a_2, a_3, ... a_n. Hence, by the above Lemma, it must be possible to form a mutually reconcileable set from the system x, b_1, a_2, a_3, ... a_n. Further, of this set x must be one; for otherwise we should have b_1 reconcileable with one or more of the system a_2, a_3, ... a_n, and also with a_1 (combined, it may be, with others of the system). Hence, by the Lemma, it would be possible to form a reconcileable set from the system a_1, a_2, ... a_n, contrary to hypothesis. Hence x must be reconcileable with one or more of the system b_1, a_2, a_3, ... a_n. If a_1 be *not* included in the set of a's with which x is reconcileable this result is obvious. Hence, in any case, we

---

* Every circuit spoken of in this note is supposed to be simple, and non-evanescible.

can, in the system a_1, a_2, ... a_n, replace a_1 by b_1, and still obtain a system of n circuits with one or more of which every other circuit drawn in the region is reconcileable.

In particular, b_2 is reconcileable with one or more of the system b_1, a_2, a_3, ... a_n; and it is, by hypothesis, not reconcileable with b_1 alone. If then a_2 be one of the set with which it is reconcileable, it follows by exactly the same argument as before that we may replace a_2 by b_2, so that any circuit drawn in the region is reconcileable with one or more of the system b_1, b_2, a_3, ... a_n.

This process of replacing a's by b's may be continued, until we finally arrive at the result that every circuit drawn in the region is reconcileable with one or more of the system b_1, b_2, ... b_n.

The order of connection of a region is therefore a perfectly definite number. For let there be *two* sets, a_1, a_2, ... a_m, and b_1, b_2, ... b_n, of independent circuits, such that any other circuit drawn in the region is reconcileable with one or more of either set; and if possible let m, n be unequal, say $n > m$. The above argument shews that every circuit drawn in the region is reconcileable with one or more of the system b_1, b_2, ... b_m, contrary to the supposition that there are $n - m$ independent circuits b_{m+1}, ... b_n which are not reconcileable.

It is possible to draw a barrier meeting any one of the circuits a_1, a_2, ... a_n in one point only. For if every barrier through a point P of one of these circuits meets the circuit again, it must be possible to connect P with all other points of the circuit by a continuous series of lines drawn each on a barrier, and therefore to fill up the circuit by a continuous surface lying wholly within the region, contrary to the suppositions that the circuit is simple and non-evanescible.

Further, it is possible to draw the barrier so as not to meet any of the remaining circuits. For if every barrier which meets one of the circuits, say a_1, necessarily intersects one at least of a certain set a of the remaining circuits, then, if we consider the infinite variety of barriers which can be so drawn, it appears that we can connect the various points of a_1 with those of a by a continuous web of lines drawn each on a barrier, and therefore we can connect a_1 and a by a continuous surface lying wholly within the region. That is, a_1 and a are reconcileable contrary to hypothesis.

The above demonstrations are slightly modified from Riemann [*] and Königsberger [†].

---

[*] *l. c.* Art. 63.

[†] *Theorie der elliptischen Functionen*, c. 5.

Let an infinitely great force X act for an infinitely short time T on the solid in the direction of x, so as to change the x-component of the impulse from ξ to $\xi + \delta\xi$. The total work done by this force is

$$\int_0^T Xu\,dt.$$

If u_1, u_2 be the greatest and least values of u during the time T, this lies between

$$u_1 \int_0^T X\,dt, \quad \text{and} \quad u_2 \int_0^T X\,dt,$$

i.e. between $u_1\delta\xi$, and $u_2\delta\xi$. If $\delta\xi$ be infinitely small, u_1, u_2 are each sensibly equal to u, and the work done is $u\delta\xi$. In the same way we may calculate the work necessary to increase the remaining components of the impulse by the infinitely small amounts $\delta\eta$, $\delta\zeta$, $\delta\lambda$, &c. Since the work done is equal to the additional kinetic energy generated, we have

$$u\delta\xi + \ldots + p\delta\lambda + \ldots = \delta T$$

$$= \frac{dT}{du}\delta u + \ldots + \frac{dT}{dp}\delta p + \ldots\ldots\ldots\ldots\ldots(1).$$

Now if the motion of the solid be altered in any ratio, the impulse will be altered in the same ratio. Let us then take

$$\frac{\delta u}{u} = \ldots = \frac{\delta p}{p} = \ldots,$$

and therefore

$$\frac{\delta\xi}{\xi} = \ldots = \frac{\delta\lambda}{\lambda} = \ldots.$$

We have, from (1),

$$u\xi + \ldots + p\lambda + \ldots = \frac{dT}{du}u + \ldots + \frac{dT}{dp}p + \ldots,$$

$$= 2T \ldots\ldots\ldots\ldots\ldots\ldots\ldots\ldots\ldots(2),$$

since T is a homogeneous quadratic function. Performing the arbitrary variation δ on both sides of (2) we find

$$u\delta\xi + \ldots + p\delta\lambda + \ldots$$
$$+ \xi\delta u + \ldots + \lambda\delta p + \ldots = 2\delta T,$$

and thence, making use of (1),

$$\xi\delta u + \ldots + \lambda\delta p + \ldots = \delta T$$

$$= \frac{dT}{du}\delta u + \ldots + \frac{dT}{dp}\delta p + \ldots.$$

Since the variations δu, δp, &c. are here independent, we have

$$\xi = \frac{dT}{du}, \ldots, \quad \lambda = \frac{dT}{dp}, \ldots,$$

the formulæ to be proved.

The dynamical equations of Art. 6 may be written, when a force-potential V exists, in the form

$$\frac{du}{dt} - 2v\zeta + 2w\eta = -\frac{dP}{dx}$$

$$\frac{dv}{dt} - 2w\xi + 2u\zeta = -\frac{dP}{dy} \qquad \dots\dots\dots\dots(1),$$

$$\frac{dw}{dt} - 2u\eta + 2v\xi = -\frac{dP}{dz}$$

where
$$P = \int \frac{dp}{\rho} + V + \tfrac{1}{2}(\text{vel.})^2.$$

Differentiating the third of equations (1) with respect to y, the second with respect to z, and subtracting, we find

$$\frac{d\xi}{dt} + v\frac{d\xi}{dy} + w\frac{d\xi}{dz} - u\left(\frac{d\eta}{dy} + \frac{d\zeta}{dz}\right) = \eta\frac{du}{dy} + \zeta\frac{du}{dz} - \xi\left(\frac{dv}{dy} + \frac{dw}{dz}\right)\dots(2).$$

Remembering the relation (1) of Art. 127, and making use of the equation of continuity [(7) of Art. 8],

$$\frac{\partial\rho}{\partial t} + \rho\theta = 0,$$

we may put (2) in the form

$$\frac{\partial\xi}{\partial t} = \xi\frac{du}{dx} + \eta\frac{du}{dy} + \zeta\frac{du}{dz} + \frac{1}{\rho}\frac{\partial\rho}{\partial t},$$

or
$$\frac{\partial}{\partial t}\frac{\xi}{\rho} = \frac{\xi}{\rho}\frac{du}{dx} + \frac{\eta}{\rho}\frac{du}{dy} + \frac{\zeta}{\rho}\frac{du}{dz} \qquad \dots\dots\dots\dots(3).$$

In the same way we obtain two similar equations, which may be written down from symmetry.

Let the projections on the co-ordinate axes of the line joining two neighbouring particles on the same vortex-line be

$$\frac{\xi}{\rho}\epsilon, \quad \frac{\eta}{\rho}\epsilon, \quad \frac{\zeta}{\rho}\epsilon,$$

L. 16

respectively. The rate at which the projection on x is increasing is equal to the difference of the values of u at the two ends, *i.e.* to

$$\frac{\xi\epsilon}{\rho}\frac{du}{dx} + \frac{\eta\epsilon}{\rho}\frac{du}{dy} + \frac{\zeta\epsilon}{\rho}\frac{du}{dz},$$

or, by (3), to

$$\epsilon\frac{\partial}{\partial t}\frac{\xi}{\rho}.$$

Hence the projections of the line in question will at the time $t + dt$ be

$$\epsilon\left(\frac{\xi}{\rho} + \frac{\partial}{\partial t}\frac{\xi}{\rho}\cdot dt\right), \quad \epsilon\left(\frac{\eta}{\rho} + \frac{\partial}{\partial t}\frac{\eta}{\rho}\cdot dt\right), \quad \epsilon\left(\frac{\zeta}{\rho} + \frac{\partial}{\partial t}\frac{\zeta}{\rho}\cdot dt\right),$$

respectively, that is, the line still forms part of a vortex-line.

Also, if ds be the length of this line at any instant, we have

$$ds = \epsilon\frac{\omega}{\rho},$$

where ω is the angular velocity of the fluid. But if σ be the section of a vortex-filament having ds as axis, the product $\rho\sigma ds$ is constant with regard to the time. Hence the strength $\omega\sigma$ of the vortex is constant.

This extension of Helmholtz' proof, which is limited to the particular case $\theta = 0$, is due to Prof. Nanson.

The connection between the above proof and that by Thomson given in the text may be shewn as follows[*].

If A, B, C be the projections on the co-ordinate planes of the area of a circuit moving with the fluid, we have

$$\frac{\partial A}{\partial t} = \int(v\,dz - w\,dy),$$

where the integration is taken right round the circuit. If the circuit be infinitely small, this gives

$$\frac{\partial A}{\partial t} = A\left(\frac{dv}{dy} + \frac{dw}{dz}\right) - B\frac{dv}{dx} - C\frac{dw}{dx}$$

$$= A\theta - A\frac{du}{dx} - B\frac{dv}{dy} - C\frac{dw}{dz}\dots\dots\dots\dots\dots(4),$$

where θ has the same meaning as in the text. But (Art. 39) the circulation in the circuit is equal to $2(\xi A + \eta B + \zeta C)$, and since this is constant with regard to the time we have

$$\frac{\partial}{\partial t}(\xi A + \eta B + \zeta C) = 0\dots\dots\dots\dots\dots\dots(5);$$

---

[*] *Messenger of Mathematics*, July, 1877.

i.e. by (4) and similar equations

$$A \left(\frac{\partial \xi}{\partial t} - \xi \frac{du}{dx} - \eta \frac{du}{dy} - \zeta \frac{du}{dz} \right)$$

$$+ B \left(\frac{\partial \eta}{\partial t} - \xi \frac{dv}{dx} - \eta \frac{dv}{dy} - \zeta \frac{dv}{dz} \right)$$

$$+ C \left(\frac{\partial \zeta}{\partial t} - \xi \frac{dw}{dx} - \eta \frac{dw}{dy} - \zeta \frac{dw}{dz} \right)$$

$$+ (A\xi + B\eta + C\zeta) \theta = 0.$$

This being true for any small circuit whatever, the coefficients of A, B, C on the left-hand side must separately vanish. The equations thus obtained are equivalent to (3).

If we introduce the Lagrangian notation of Arts. 16, &c., and apply (5) to the circuit which initially bounded the rectangle $db\,dc$, we have

$$A = \frac{d(y, z)}{d(b, c)} db\,dc, \quad \&c.,$$

and therefore

$$\xi \frac{d(y, z)}{d(b, c)} + \eta \frac{d(z, x)}{d(b, c)} + \zeta \frac{d(x, y)}{d(b, c)} = \xi_0,$$

where ξ_0 is the initial value of ξ at (a, b, c). This and two similar equations which may be written down from symmetry are Cauchy's integrals of the Lagrangian equations. These integrals contain, as Stokes pointed out, the first rigorous proof of Lagrange's theorem. See Art. 23. The remark that they follow in the above manner from Thomson's circulation-theorem is due to Prof. Nanson.

If we eliminate p between the equations of motion of a viscous liquid [(15) of Art. 180], we obtain three equations of the form

$$\frac{\partial \xi}{\partial t} = \xi \frac{du}{dx} + \eta \frac{du}{dy} + \zeta \frac{du}{dz} + \mu' \nabla^2 \xi \dots\dots\dots (4).$$

The first three terms on the right-hand side express, as we have seen, the rate at which ξ varies for a particular particle when the vortex-lines move with the fluid and the strengths of all the vortices remain constant. The additional variation of ξ due to viscosity is given by the last term of (4), and follows the same law as the variation of temperature due to conduction of heat in a medium of uniform conductivity. It appears from this analogy that vortex-motion cannot originate in the interior of a viscous fluid, but must be diffused inwards from the boundary.

NOTE E.

On the Resistance of Fluids.

One of the most important but at the same time most difficult practical questions connected with the subject of this book is the determination of the resistance experienced by a solid moving through a fluid, *e.g.* by a ship moving through the water, or by a projectile through the air.

The effect of a liquid on the motion of a solid through it has been discussed, on certain assumptions, in Arts. 105—119. It was there found that the whole effect was simply equivalent to an increase in the *inertia* of the solid. The latter yields more sluggishly to the action of force than it would do if the fluid were removed, whether the tendency of the force be to increase or to diminish the motion which the body already has; but there is no tendency to a total transfer of energy of motion from the solid to the fluid, or in any other way to reduce the solid to rest. Thus a sphere immersed in a liquid will move under the action of any forces exactly as if its inertia were increased by half that of the fluid displaced, and the fluid then annihilated. For instance, if set in motion and then left to itself, it will describe a straight path with uniform velocity.

These theoretical conclusions do not at all correspond with what is observed in actual cases. As a matter of fact a body moving through a fluid requires a continual application of force to maintain the motion, and if this be not supplied the body is quickly brought to rest.

This discrepancy between theory and observation must be due to unreality in one or more of the fundamental assumptions on which the investigations referred to were based. These assumptions are

A. That the fluid is 'perfect,' *i.e.* that there is no tangential action between adjacent portions of fluid, or between the fluid and the solid;

B. That the motion is continuous, *i.e.* that the velocities u, v, w are everywhere continuous functions of the co-ordinates; and

C. That the fluid is unlimited except by the surface of the solid.

Let us take these in order.

A.   The effects of imperfect fluidity, or viscosity, have been treated of in Chapter IX.   The resistance due to this cause is proportional to the velocity of the solid, and, for a body of given shape, to its linear dimensions.   This has been proved in the text for the case of the sphere, and it is easy, by the method of 'dimensions,' to extend the result to the general case.   Thus if l, L, t, T, u, U, p, P be corresponding lengths, times, velocities, and pressures in two geometrically similar cases of motion, it appears from equations (14) of Art. 180 that we must have

$$\frac{p}{l} : \frac{P}{L} = \frac{\mu u}{l^2} : \frac{\mu U}{L^2} = \frac{\rho u}{t} : \frac{\rho U}{T}.$$

Hence

$$\frac{l^2}{t} = \frac{L^2}{T},$$

and

$$pl^2 : PL^2 = \frac{l^2 u}{t} : \frac{L^2 U}{T} = lu : LU.$$

That is, the resultants of the pressures (normal and tangential) on any corresponding areas are proportional to the products of corresponding lines and velocities.

It must be remembered however that the investigations of Chapter IX. proceed on the assumption that the motion is slow.   Thus it is very doubtful whether the equations of Art. 180, or the boundary conditions of Art. 181, would be applicable to the motion of the stratum of water in contact with the side of a ship in rapid motion.

B.   It appears from Art. 30 that there is a certain limit to the velocity of a solid of given shape if the motion be continuous.   When this limit is reached, the pressure at some point of the surface of the solid sinks to zero, and if the limit be exceeded, a surface of discontinuity is formed.   See Art. 94.   If the surface of the solid have a sharp projecting edge or angle, this limiting velocity is very small (zero if the edge be of perfect geometrical sharpness), whilst for bodies of 'fair' easy shape it may be considerable.

When the motion is continuous a certain amount of momentum is expended in each unit of time in starting the elementary streams which diverge from the body in front, but this momentum is restored again to the body by the streams as they close in behind.   But when a surface of discontinuity is formed the streams do not close in again; on the contrary, we have a mass of 'dead water' following the body and pressing on its rear with merely the general pressure which obtains in the parts of the fluid which are at rest. Hence the restoration of momentum no longer takes place, and a force equal to the loss per unit time must be applied to the solid in order to maintain its motion.

The only case of this kind which has as yet been mathematically worked out is that in which the solid is a long plane lamina, with parallel straight edges. If the lamina move broadside-on, the motion is obtained from Art. 98 by considering a velocity q_0 parallel to y impressed on everything. The resistance to the motion of the lamina with velocity q_0 is then $\frac{\pi}{4+\pi} \rho q_0{}^2 l$, where l is the breadth of the lamina. The resistance experienced by a lamina moving obliquely has been calculated by Lord Rayleigh (*l. c.* Art. 96). The result is

$$\frac{\pi \cos a}{4 + \pi \cos a} \rho q_0{}^2 l,$$

where a is the angle which the direction of motion of the lamina makes with the normal to its plane.

Generally the resistance due to this cause is proportional to the square of the velocity, so long as we can assume that the motions of the fluid corresponding to different velocities of the solid are geometrically similar.

The formation of the 'wake' which is observed to follow a vessel in motion has been attributed by some writers to the friction between the surface of the ship and the water. It is supposed that the tendency of this friction is to drag a mass of water bodily after the ship. This explanation, if correct, shews that the laws of fluid friction laid down in Chapter IX. do not hold for such rapid motions as are here in question.

It seems possible however that the wake may be in a great measure due to the cause at present under consideration. It is at all events admitted that the wake is greatly increased by discontinuity in the lines of the vessel, a state of things favourable as we have seen to the formation of a surface of discontinuity in the fluid. Again* it is possible that the bottom of a ship (especially when foul) is to be re-garded not as a geometrical surface, but as a mass of projections each of which establishes a surface of discontinuity, with dead water in its rear, on its own account. The aggregate of these masses of dead water would then build up the wake. We should thus also have an explanation of the law, laid down by some writers on this subject, that the 'skin-resistance' is proportional to the square of the velocity.

* Cf. Stokes, *Camb. Trans.*, vol. VIII, p. 301.

Of course there is not, practically, in any of these cases absolute discontinuity of motion. It was shewn in Art. 132 that a surface of discontinuity may be regarded as made up of a system of vortices distributed in a certain way. Owing to fluid friction the vortex-motion does not remain concentrated in this surface, but is diffused through the fluid on each side. Thus the boundary of the dead water is marked by a band of eddies.

C. The effect of a rigid boundary to the fluid may be estimated from Art. 125 (a). It appears that if the dimensions of the solid be small compared with its distance from the boundary, the influence of the latter may be neglected.

A free surface at a great distance from it also has little effect on a solid. The case is otherwise however when the solid is only partially submerged, e.g. a ship. One effect of the motion is to make the pressure deviate from its statical value; to increase it, for instance, at the bows and at the stern, and to diminish it amidships[*]. Hence the level of the fluid is disturbed, there is an elevation of the water at each of the former points, and a depression between. A wave like this, accompanying the ship, would of course cause no loss of energy beyond what is necessary to maintain it against viscosity. But we have also waves produced which travel over the surface, and carry off energy to the distant parts of the fluid. The energy thus dispersed must of course come directly or indirectly from the ship; and the loss from this cause constitutes a special form of resistance, called 'wave-resistance[†].'

A body moving through air will in like manner experience a resistance due to the dispersion of energy by air-waves. It appears, however, that if the motion be steady, and the velocity of the body small compared with that of sound, the resistance due to this cause is inappreciable. For we have, with the same notation as in Art. 169,

$$ s = -\frac{1}{c^2}\frac{d\phi}{dt}, \qquad \nabla^2\phi = -\frac{ds}{dt}; $$

[*] This is easily seen by impressing on everything a velocity equal and opposite to that of the ship, and so reducing the case to one of steady motion.

[†] For a discussion of the various kinds of waves produced by a ship, and of the probable laws of wave-resistance, we must refer to a paper by Rankine, in the *Phil. Trans.* for 1871. The student may also consult, on the general question, Rankine "On Stream-Lines in connection with Naval Architecture," *Nature* vol. II, and Froude "On Stream-Lines in Relation to the Resistance of Ships," *Nature*, vol. III.

and writing $\phi = a\psi$, where a is the velocity of the solid, we have, as in Arts. 86, 105, to write for $\dfrac{d}{dt}$, $-a\dfrac{d}{dx}$, when the quantities to be differentiated are referred to axes fixed in the solid. On this supposition the above formulæ become

$$s = \frac{a}{c^2}\frac{d\psi}{dx}, \qquad \nabla^2\phi = a\frac{ds}{dx},$$

and therefore

$$\nabla^2\phi = \frac{a^2}{c^2}\frac{d^2\psi}{dx^2},$$

i.e. $\nabla^2\phi$ is of the second order of small quantities. Hence to a first approximation

$$\nabla^2\phi = 0,$$

i.e. the fluid moves sensibly as if it were incompressible.

LIST OF MEMOIRS AND TREATISES.

MEMOIRS.

EULER, L. Principes généraux du mouvement des fluides. *Hist. de l'Acad. de Berlin.* 1755.

„ De principiis motus fluidorum. *Novi Comm. Acad. Petrop.* t. 14, p. 1. 1759.

LAGRANGE, J. L. Mémoire sur la théorie du mouvement des fluides. *Nouv. Mém. de l'Acad. de Berlin.* 1781.

POISSON, S. D. Sur les équations générales de l'équilibre et du mouvement des corps solides élastiques et des fluides. *Journ. de l'Ecole Polyt.* t. 13. 1819.

GREEN, G. Researches on the vibrations of pendulums in fluid media. *Trans. R. S. Edin.* Vol. 13. 1833.

„ On the motion of waves in a variable canal of small depth and width. *Camb. Trans.,* vol. 6, 1837.

„ Note on the motion of waves in canals. *Camb. Trans.,* vol. 7. 1839.

STOKES, G. G. On the steady motion of incompressible fluids. *Camb. Trans.,* vol. 7. 1842.

„ On some cases of fluid motion. *Camb. Trans.,* vol. 8. 1843.

SCOTT RUSSELL, J. Report on waves. *B. A. Reports.* 1844.

AIRY, G. B. Tides and waves. *Encyc. Metrop.,* vol. 5. 1845.

STOKES, G. G. On the theories of the internal friction of fluids in motion, and of the equilibrium and motion of elastic solids. *Camb. Trans.,* vol. 8. 1845.

„ Report on recent researches in hydrodynamics. *B. A. Reports.* 1846.

„ Supplement to a memoir on some cases of fluid motion. *Camb. Trans.,* vol. 8. 1846.

„ On the theory of oscillatory waves. *Camb. Trans.,* vol. 8. 1847.

THOMSON, W. Notes on hydrodynamics. *Camb. and Dub. Math. J.*, vol. 4. 1849.

STOKES, G. G. On waves. *Camb. and Dub. Math. J.*, vol. 4. 1849.

„ On the effect of the internal friction of fluids on the motion of pendulums. *Camb. Trans.*, vol. 9. 1851.

CLEBSCH, A. Ueber die Bewegung eines Ellipsoids in einer tropfbaren Flüssigkeit. *Crelle*, t. 52. 1856; and t. 53. 1857.

HELMHOLTZ, H. Ueber Integrale der hydrodynamischen Gleichungen welche den Wirbelbewegungen entsprechen. *Crelle*, t. 55. 1858.

RIEMANN, B. Ueber die Fortpflanzung ebener Luftwellen von endlicher Schwingungsweite. *Gött. Abh.*, t. 8. 1860.

DIRICHLET, P. LE-J. Untersuchungen über ein Problem der Hydrodynamik. *Gött. Abh.*, t. 8. 1860.

EARNSHAW, S. On the mathematical theory of sound. *Phil. Trans.* 1860.

HELMHOLTZ and PIOTROWSKI. Ueber Reibung tropfbarer Flüssigkeiten. *Wien. Sitz. B.* t. 40. 1860.

MEYER, O. E. Ueber die Reibung der Flüssigkeiten. *Crelle*, t. 59. 1861.

RIEMANN, B. Beitrag zu den Untersuchungen über die Bewegung eines flüssigen gleichartigen Ellipsoids. *Gött. Abh.*, t. 9. 1861.

RANKINE, W. J. M. On the exact form of waves near the surface of deep water. *Phil. Trans.* 1863.

STEFAN, M. J. Ueber die Bewegung flüssiger Körper. *Wien. Sitz. B.* t. 46. 1863.

RANKINE, W. J. M. On plane water-lines in two dimensions. *Phil. Trans.* 1864.

„ Summary of the properties of certain stream-lines. *Phil. Mag.* 1864.

MAXWELL, J. C. On the viscosity or internal friction of air and other gases. *Phil. Trans.* 1866.

THOMSON, W. On vortex atoms. *Phil. Mag.* 1867.

RANKINE, W. J. M. On waves in liquids. *Proc. R. S.* 1868.

HELMHOLTZ, H. Ueber discontinuirliche Flüssigkeitsbewegungen. *Berl. Monatsb.* 1868.

WEBER, H. Ueber eine Transformation der hydrodynamischen Gleichungen. *Crelle*, t. 68. 1868.

RANKINE, W. J. M. On the thermodynamic theory of waves of finite longitudinal disturbance. *Phil. Trans.* 1870.

THOMSON, W. On vortex motion. *Trans. R. S. Edin.*, vol. 25. 1869.

KIRCHHOFF, G.   Zur Theorie freier Flüssigkeitsstrahlen.   *Crelle*, t. 70. 1869.

„   Ueber die Bewegung eines Rotationskörpers in einer Flüssigkeit. *Crelle*, t. 71. 1870.

„   Ueber die Kräfte, welche zwei unendlich dünne, starre Ringe in einer Flüssigkeit scheinbar auf einander ausüben können. *Crelle*, t. 71. 1870.

MAXWELL, J. C.   On the displacement in a case of fluid motion.   *Proc. Lond. Math. Soc.*, vol. 3. 1870.

MEYER, O. E.   Ueber die pendelnde Bewegung einer Kugel unter dem Einflusse der inneren Reibung des umgebenden Mediums. *Crelle*, t. 73. 1871.

RANKINE, W. J. M.   On the mathematical theory of stream-lines, especially those with four foci and upwards. *Phil. Trans.* 1871.

THOMSON, W. Hydrokinetic solutions and observations. *Phil. Mag.* 1871.

„   On the ultra-mundane corpuscles of Le Sage, also on the motion of rigid solids in a liquid circulating irrotationally through perforations in them or in a fixed solid.   *Phil. Mag.* 1873.

RAYLEIGH, LORD.   On waves.   *Phil. Mag.* 1876.

„   On the resistance of fluids.   *Phil. Mag.* 1877.

„   Notes on hydrodynamics.   *Phil. Mag.* 1877.

GREENHILL, A. G.   Plane vortex motion.   *Quart. J. of Math.*, vol. 15. 1877.

TREATISES.

LAGRANGE, J. L.   Mécanique analytique.   Ed. 1815, Part 2, Sect. 10, 11, 12.

POISSON, S. D.   Traité de mécanique.   2me éd.   Paris, 1833.   T. 2, liv. 6.

THOMSON, W. and TAIT, P. G.   Treatise on natural philosophy.   Oxford, 1867. §§ 331—336. New edition: Cambridge, 1879. §§ 320—330.

RIEMANN, B.   Partielle Differentialgleichungen und deren Anwendung auf physikalische Fragen.   Hrsg. v. Hattendorff.   Braunschweig, 1869. §§ 98—109.

RÉSAL, H.   Traité de mécanique générale.   Paris, 1874.   T. 2, cc. 13, 14.

KIRCHHOFF, G.   Vorlesungen über mathematische Physik.   Leipzig, 1874—6. Cc. 15—26.

BESANT, W. H.   Treatise on Hydromechanics.   3rd edition, Cambridge, 1877. Cc. 10—15.

EXERCISES.

1. If the motion of a fluid in two dimensions be referred to polar co-ordinates r, θ, and if u, v denote the component velocities along and perpendicular to the radius vector, the component accelerations in the same directions are

$$\frac{du}{dt} + u\frac{du}{dr} + v\frac{du}{rd\theta} - \frac{v^2}{r},$$

and

$$\frac{dv}{dt} + u\frac{dv}{dr} + v\frac{dv}{rd\theta} + \frac{uv}{r},$$

respectively.

2. If in the solution of the equations

$$\frac{dx}{dt} = u, \qquad \frac{dy}{dt} = v, \qquad \frac{dz}{dt} = w,$$

where u, v, w are given functions of x, y, z, t, satisfying the condition

$$\frac{du}{dx} + \frac{dv}{dy} + \frac{dw}{dz} = 0,$$

x, y, z be expressed as functions of t, and the arbitrary constants a, b, c, the determinant $\dfrac{d(x, y, z)}{d(a, b, c)}$ is independent of t.

3. If $F(x, y, z, t) = 0$ be the equation of a moving surface, the velocity of the surface normal to itself at any point is

$$-\frac{1}{R}\frac{dF}{dt}, \quad \text{where } R^2 = \left(\frac{dF}{dx}\right)^2 + \left(\frac{dF}{dy}\right)^2 + \left(\frac{dF}{dz}\right)^2.$$

Hence deduce the surface-condition of Art. 10.　　　　[Thomson.]

4.   In the case of motion in two dimensions for which

$$udx + vdy = \tfrac{1}{2}d\,(y^2 - x^2),$$

apply (10) of Art. 10 to obtain the general equation of lines made up of the same particles; and thence shew that the particles which once lie in a curve of the nth order continue to lie in a curve of the nth order.

[Stokes.]

5.   Investigate an expression for the change in an indefinitely short time in the mass of fluid contained within a spherical surface of small radius.

Prove that the momentum of the mass in the direction of the axis of x is greater than it would be if the whole were moving with the velocity at the centre by

$$\frac{1}{5}\,\frac{Ma^2}{\rho}\left\{\frac{d\rho}{dx}\frac{du}{dx} + \frac{d\rho}{dy}\frac{du}{dy} + \frac{d\rho}{dz}\frac{du}{dz} + \tfrac{1}{2}\rho\left(\frac{d^2u}{dx^2} + \frac{d^2u}{dy^2} + \frac{d^2u}{dz^2}\right)\right\}.$$

[H. M. Taylor, *Math. Trip.*, 1876.]

6.   A cistern discharges water into the atmosphere through a vertical pipe of uniform section. Shew that air would be sucked in through a small hole in the upper part of the pipe, and explain how this result is consistent with an atmospheric pressure in the cistern.

[Lord Rayleigh, *Math. Trip.*, 1876.]

7.   Let a spherical portion of an infinite quiescent liquid be separated from the liquid round it by an infinitely thin flexible membrane, and let this membrane be suddenly set in motion, every part of it in the direction of the radius and with velocity equal to S_i, a harmonic function of position on the surface. Find the velocity produced at any external or internal point of the liquid.

[Thomson.]

8.   Prove that the energy of the irrotational motion of a liquid in a given region is less than that of any other continuous motion consistent with the same motion of the boundary.

[Thomson.]

9.   Prove that if the force-potential V satisfy the relation $\nabla^2 V = 0$ the pressure cannot, in irrotational motion, be a minimum at any point in the interior of an incompressible fluid.

10.   In the irrotational motion of a fluid in two dimensions prove that if the velocity be everywhere the same in magnitude it is so in direction.

[*Math. Trip.*, 1873.]

11. Prove, and interpret, the following formula for the energy of a liquid moving irrotationally in two dimensions: $2T = \rho \int \phi\, d\psi$, the integration extending round the boundary.

12. When a liquid moves irrotationally in two dimensions round a corner where two branches of the boundary meet at an angle α (measured on the side of the liquid) which is $< 180°$, shew that the particle at the corner is in a position of permanent or instantaneous rest according as $\alpha \lessgtr 90°$ or $> 90°$.

What takes place when $\alpha > 180°$? [Stokes.]

13. A stream of uniform depth and of uniform width $2a$ flows slowly through a bridge consisting of two equal arches resting on a rectangular pier of width $2b$, the bridge being so broad that under it the fluid moves uniformly with velocity U. Shew that, after the stream has passed through the bridge, the velocity-potential of the motion is

$$\frac{a-b}{a} Ux + \frac{2aU}{\pi^2} \Sigma_1^\infty \frac{1}{n^2} \sin \frac{n\pi b}{a} \cos \frac{n\pi y}{a} e^{-\frac{n\pi x}{a}},$$

the axis of x being in the forward direction of the stream, and the origin at the middle point of the pier.

Find the equation of the path of any particle of the water.

[C. Niven, *Math. Trip.*, 1878.]

14. The transverse section of a uniform prismatic closed vessel is of the form bounded by two intersecting hyperbolas represented by the equations $\sqrt{2}(x^2 - y^2) + x^2 + y^2 = a^2$, $\sqrt{2}(y^2 - x^2) + x^2 + y^2 = b^2$.

If the vessel be filled with water, and be made to rotate with angular velocity ω about its axis, prove that the initial component velocities of any point (x, y) of the water will be

$$\frac{\omega}{a^2 + b^2} \{2y^3 - 6x^2 y + \sqrt{2}(a^2 - b^2) y\},$$

and

$$-\frac{\omega}{a^2 + b^2} \{2x^3 - 6xy^2 + \sqrt{2}(a^2 - b^2) x\},$$

respectively. [Ferrers, *Math. Trip.*, 1872.]

15. Work out by the method of Art. 95 the solution in the case where a stream of liquid, of given breadth, impinges perpendicularly on an infinite plane lamina. [Kirchhoff.]

16. An anchor-ring is in motion parallel to its axis in an infinite mass of liquid. Shew by a diagram the arrangement of the stream-lines and equipotential surfaces, firstly when there is, and secondly when there is not, cyclic motion through the ring.

17. A large closed vessel of incompressible liquid with a motionless solid globe immersed in it at a distance from the nearest part of the outer boundary very great in comparison with the radius of the globe, is suddenly set in motion with a given translational velocity V. Find the instantaneous motion of the liquid and of the globe. Prove that according as the mean density of the globe is greater or less than the density of the liquid, its velocity is less or greater than V. (Assume the centre of inertia of the globe to be in its centre of figure.)

[Thomson.]

18. Prove that under the circumstances of Arts. 107, 108 the linear momentum of the whole matter contained within a spherical surface enclosing the whole of the moving solids is equal to two-thirds of the impulse; also that the linear momentum of the matter contained within a spherical surface not enclosing any of the solids is the same as if this matter were rigid and moving with the velocity at the centre.

[Thomson.]

19. Verify the former of the statements in the preceding exercise in the case of the sphere (Art. 105).

20. A sphere is in motion in a mass of liquid enclosed in a fixed spherical case. Find the motion of the liquid at the instant when the centre of the sphere coincides with that of the case. [Stokes.]

21. Also work out, by the method of Art. 125, an approximate solution for the case in which the distance between the centres is small compared with the radius of the containing vessel.

22. Integrate the equations of motion of a solid through a liquid in the case of Art. 116 (g).

[Thomson.]

23. Work out the problem of Art. 125, supposing the spheres to be moving perpendicularly to the line of their centres.

24. Prove (by the method of images) that a straight vertical vortex-filament moving in water bounded by two vertical walls at right angles will describe the Cotes's spiral $r \sin 2\theta = a$ in exactly the same manner as a particle under the action of a repulsion varying inversely as the cube of the distance.

Prove also by the same method that if the vertical planes be inclined at an angle $\dfrac{\pi}{n}$, an exact submultiple of two right angles, the vortex-filament will describe similarly the Cotes's spiral $r \sin n\theta = a$.

[Greenhill.]

25.　If (r_1, θ_1), (r_2, θ_2), ... be the polar co-ordinates at time t of a system of rectilinear vortices whose strengths are m_1, m_2, ... respectively, prove that

$$\Sigma mr^2 = \text{const.,}$$

and

$$\Sigma mr^2 \frac{d\theta}{dt} = \frac{1}{\pi} \Sigma m_1 m_2.$$

[Kirchhoff.]

26.　A series of long waves is propagated along a uniform canal, arising from a small disturbance at the mouth which varies as a given function of the time; assuming that the effect of friction may be represented in a general way by a retarding force varying as the velocity, determine the motion.　　　　　　　　　　　　　　[Stokes.]

27.　Prove that the formula

$$\phi = a \left(e^{jy} + e^{-jy}\right) \left(e^{jz} + e^{-jz}\right) \cos k \left(x - ct\right),$$

where $j = \dfrac{k}{\sqrt{2}}$, represents a possible form of wave motion in a uniform canal whose section is a right-angled isosceles triangle having its sides equally inclined to the vertical.　(The axis of x is the bottom line of the canal, and that of y is vertical.)

Find the velocity of propagation in terms of the wave-length; and examine the form of the free surface.　　　　　　　　[Kelland.]

28.　Prove the formula

$$\phi = ae^{k(y \sin \beta - z \cos \beta)} \cos k \left(x - ct\right),$$

giving the motion of a series of waves parallel to the edge of a shore sloping at an angle β.

Find the velocity of propagation, and the form of the free surface.
　　　　　　　　　　　　　　　　　　[Stokes.]

29.　Calculate the tidal motion of a heavy liquid contained in a square vessel of uniform depth, due to a small horizontal disturbing force acting uniformly throughout the mass, whose magnitude is constant, and whose direction revolves uniformly in the horizontal plane.

How could the forces here imagined be realized experimentally?
　　　　　　　　　　　[Lord Rayleigh, *Math. Trip.*, 1876.]

30. If liquids of densities ρ and ρ' and depths h and h' be contained between two fixed horizontal planes at a distance $h + h'$, prove that the velocity v of propagation of waves of small displacement of length λ at the common surface is given by

$$(v - V \cos a)^2 \rho \coth \frac{2\pi h}{\lambda} + (v - V' \cos a')^2 \rho' \coth \frac{2\pi h'}{\lambda} - \frac{g\lambda}{2\pi}(\rho - \rho') = 0,$$

where V and V' are the mean velocities of the currents in the liquids, and a and a' are the angles the currents make with the direction of propagation of the waves, the currents slipping over each other.

[Greenhill, *Math. Trip.*, 1878.]

31. A wind of velocity V is blowing horizontally over the surface of deep water; prove that the velocity of propagation of waves of length λ is

$$\frac{\sigma V}{1 + \sigma} \pm \left\{ c^2 - \frac{\sigma V^2}{(1 + \sigma)^2} \right\}^{\frac{1}{2}},$$

where $\qquad c = \left\{ \frac{g\lambda}{2\pi} \cdot \frac{1 - \sigma}{1 + \sigma} \right\}^{\frac{1}{2}} = $ velocity without wind,

$\sigma =$ specific gravity of air, and the upper or lower sign is to be taken according as the waves travel with or against the wind.

[Thomson.]

32. Hence shew that the velocity of waves travelling with the wind is greater or less than the velocity of the same waves without wind, according as $V >$ or $< 2c$; and that waves of length λ cannot travel against the wind if

$$V > c \sqrt{\frac{1 + \sigma}{\sigma}}. \qquad\qquad \text{[Thomson.]}$$

33. Find an expression for the average energy transmitted across a fixed vertical plane parallel to the fronts of an infinite train of irrotational harmonic waves, of given small elevation, moving on water of uniform depth. [*Math. Trip.*, 1878.]

34. A deep rectangular vessel nearly filled with water is continued at one end as a shallow canal of indefinite length; supposing the water of the vessel thrown into the condition of a stationary undulation, find approximately the rate at which the undulations would subside by communication to the water of the canal. [Stokes.]

35. From the general properties of equipotential and stream-curves, prove that in a regular series of waves moving in deep water without molecular rotation there is necessarily in the neighbourhood of the surface a transference of fluid in the direction of the wave's propagation, whether that surface satisfy the condition of a free surface or not.

[Lord Rayleigh, Math. Tripos, 1876.]

36. Prove that a wave of sudden rarefaction in a gas is an unstable state of motion. [Thomson.]

37. Prove that the dissipation-function of Art. 179 cannot be zero at every point of an incompressible fluid unless the fluid move as a solid body. [Stokes.]

38. Assuming the formula

$$\phi = ae^{-ky}\sin k\,(x - ct)$$

as representing approximately the motion of a long train of waves on deep water, prove by means of the dissipation-function of Art. 179 that the diminution of a due to viscosity is given by the formula

$$a = a_0 e^{-4k^2\mu' t}.$$ [Stokes.]

39. Explain in general terms, without calculation, why short waves are more rapidly destroyed by viscosity than long ones.

40. A long circular cylinder rotates about its axis with uniform velocity in an infinite mass of viscous fluid; find the motion of the fluid when it has become steady. [Stokes.]

41. Work out the same problem for the case of a rotating sphere. [Kirchhoff.]

42. Investigate the small oscillations of a spherical shell filled with viscous fluid, and oscillating by the torsion of a suspending wire. [Helmholtz.]

43. A cylinder moves with uniform velocity in a direction perpendicular to its length, in an infinite mass of viscous fluid; prove that the motion cannot be steady. Describe in general terms the nature of the actual motion. [Stokes.]

CAMBRIDGE: PRINTED BY C. J. CLAY, M.A., AT THE UNIVERSITY PRESS.

CAMBRIDGE UNIVERSITY PRESS PUBLICATIONS.

THE ELECTRICAL RESEARCHES OF THE HONOUR-
ABLE HENRY CAVENDISH, F.R.S. Written between 1771 and 1781.
Edited from the original manuscripts in the possession of the Duke of
Devonshire, K.G., by J. CLERK MAXWELL, F.R.S. 8vo.

MATHEMATICAL AND PHYSICAL PAPERS. By GEORGE
GABRIEL STOKES, M.A., D.C.L., LL.D., F.R.S., Fellow of Pembroke College
and Lucasian Professor of Mathematics in the University of Cambridge.
Reprinted from the Original Journals and Transactions, with Additional Notes
by the Author. [In the Press.

A TREATISE ON NATURAL PHILOSOPHY. Vol. I., Part
I. By Sir W. THOMSON, LL.D., D.C.L., F.R.S., Professor of Natural Phi-
losophy in the University of Glasgow, and P. G. TAIT, M.A., Professor of
Natural Philosophy in the University of Edinburgh. *Second Edition.* 8vo. 16s.

ELEMENTS OF NATURAL PHILOSOPHY. By Professors
Sir W. THOMSON and P. G. TAIT. Part I. *Second Edition.* 8vo. cloth. 9s.

THE ANALYTICAL THEORY OF HEAT. By JOSEPH
FOURIER. Translated, with Notes, by A. FREEMAN, M.A., Fellow of St John's
College, Cambridge. Demy 8vo. 16s.

Cambridge:

AT THE UNIVERSITY PRESS.

London: CAMBRIDGE WAREHOUSE, 17, PATERNOSTER Row.
Cambridge: DEIGHTON, BELL, AND CO.

UNIVERSITY PRESS, CAMBRIDGE,
April, 1884.

CATALOGUE OF

WORKS

PUBLISHED FOR THE SYNDICS

OF THE

Cambridge University Press.

·

London: C. J. CLAY, M.A. AND SON.

CAMBRIDGE UNIVERSITY PRESS WAREHOUSE,

~~17 PATERNOSTER ROW.~~ AVE MARIA LANE.

GLASGOW: 263, ARGYLE STREET.

Cambridge: DEIGHTON, BELL, AND CO.

Leipzig: F. A. BROCKHAUS.

1000
2/4 84

The Cambridge University Press.

THE HOLY SCRIPTURES, &c.

THE CAMBRIDGE PARAGRAPH BIBLE of the Authorized English Version, with the Text Revised by a Collation of its Early and other Principal Editions, the Use of the Italic Type made uniform, the Marginal References remodelled, and a Critical Introduction prefixed, by F. H. A. SCRIVENER, M.A., LL.D., Editor of the Greek Testament, Codex Augiensis, &c., and one of the Revisers of the Authorized Version. Crown 4to. gilt. 21*s.*

From the Times.

"Students of the Bible should be particularly grateful to (the Cambridge University Press) for having produced, with the able assistance of Dr Scrivener, a complete critical edition of the Authorized Version of the English Bible, an edition such as, to use the words of the Editor, 'would have been executed long ago had this version been nothing more than the greatest and best known of English classics.' Falling at a time when the formal revision of this version has been undertaken by a distinguished company of scholars and divines, the publication of this edition must be considered most opportune."

From the Athenæum.

"Apart from its religious importance, the English Bible has the glory, which but few sister versions indeed can claim, of being the chief classic of the language, of having, in conjunction with Shakspeare, and in an immeasurable degree more than he, fixed the language beyond any possibility of important change. Thus the recent contributions to the literature of the subject, by such workers as Mr Francis Fry and Canon Westcott, appeal to a wide range of sympathies; and to these may now be added Dr Scrivener, well known for his labours in the cause of the Greek Testament criticism, who has brought out, for the

Syndics of the Cambridge University Press, an edition of the English Bible, according to the text of 1611, revised by a comparison with later issues on principles stated by him in his Introduction. Here he enters at length into the history of the chief editions of the version, and of such features as the marginal notes, the use of italic type, and the changes of orthography, as well as into the most interesting question as to the original texts from which our translation is produced."

From the Methodist Recorder.

"This noble quarto of over 1300 pages is in every respect worthy of editor and publishers alike. The name of the Cambridge University Press is guarantee enough for its perfection in outward form, the name of the editor is equal guarantee for the worth and accuracy of its contents. Without question, it is the best Paragraph Bible ever published, and its reduced price of a guinea brings it within reach of a large number of students."

From the London Quarterly Review.

"The work is worthy in every respect of the editor's fame, and of the Cambridge University Press. The noble English Version, to which our country and religion owe so much, was probably never presented before in so perfect a form."

THE CAMBRIDGE PARAGRAPH BIBLE. STUDENT'S EDITION, on *good writing paper*, with one column of print and wide margin to each page for MS. notes. This edition will be found of great use to those who are engaged in the task of Biblical criticism. Two Vols. Crown 4to. gilt. 31*s.* 6*d.*

THE LECTIONARY BIBLE, WITH APOCRYPHA, divided into Sections adapted to the Calendar and Tables of Lessons of 1871. Crown 8vo. 3*s.* 6*d.*

THE BOOK OF ECCLESIASTES, with Notes and Introduction. By the Very Rev. E. H. PLUMPTRE, D.D., Dean of Wells. Large Paper Edition. Demy 8vo. 7*s.* 6*d.*

"No one can say that the Old Testament is a dull or worn-out subject after reading this singularly attractive and also instructive commentary. Its wealth of literary and historical illustration surpasses anything to which we can

point in English exegesis of the Old Testament; indeed, even Delitzsch, whose pride in it is to leave no source of illustration unexplored, is far inferior on this head to Dr Plumptre."—*Academy,* Sept. 10, 1881.

London: Cambridge University Press Warehouse, 17 *Paternoster Row.*

BREVIARIUM AD USUM INSIGNIS ECCLESIAE

SARUM. Juxta Editionem maximam pro CLAUDIO CHEVALLON ET FRANCISCO REGNAULT A.D. MDXXXI. in Alma Parisiorum Academia impressam : labore ac studio FRANCISCI PROCTER, A.M., ET CHRISTOPHORI WORDSWORTH, A.M.

FASCICULUS I. In quo continentur KALENDARIUM, et ORDO TEMPORALIS sive PROPRIUM DE TEMPORE TOTIUS ANNI, una cum ordinali suo quod usitato vocabulo dicitur PICA SIVE DIRECTORIUM SACERDOTUM. Demy 8vo. 18s.

"The value of this reprint is considerable to liturgical students, who will now be able to consult in their own libraries a work absolutely indispensable to a right understanding of the history of the Prayer-Book, but which till now usually necessitated a visit to some public library, since the rarity of the volume made its cost prohibitory to all but a few. . . . Messrs Procter and Wordsworth have discharged their editorial task with much care and judgment, though the conditions under which they have been working are such as to hide that fact from all but experts."—*Literary Churchman.*

FASCICULUS II. In quo continentur PSALTERIUM, cum ordinario Officii totius hebdomadae juxta Horas Canonicas, et proprio Completorii, LITANIA, COMMUNE SANCTORUM, ORDINARIUM MISSAE CUM CANONE ET XIII MISSIS, &c. &c. Demy 8vo. 12s.

"Not only experts in liturgiology, but all persons interested in the history of the Anglican Book of Common Prayer, will be grateful to the Syndicate of the Cambridge University Press for forwarding the publication of the volume which bears the above title, and which has recently appeared under their auspices."— *Notes and Queries.*
"Cambridge has worthily taken the lead with the Breviary, which is of especial value for that part of the reform of the Prayer-Book which will fit it for the wants of our time . . . .

For all persons of religious tastes the Breviary, with its mixture of Psalm and Anthem and Prayer and Hymn, all hanging one on the other, and connected into a harmonious whole, must be deeply interesting."—*Church Quarterly Review.*
"The editors have done their work excellently, and deserve all praise for their labours in rendering what they justly call ' this most interesting Service-book' more readily accessible to historical and liturgical students."— *Saturday Review.*

FASCICULUS III. *Nearly ready.*

GREEK AND ENGLISH TESTAMENT, in parallel

Columns on the same page. Edited by J. SCHOLEFIELD, M.A. late Regius Professor of Greek in the University. Small Octavo. New Edition, with the Marginal References as arranged and revised by Dr SCRIVENER. Cloth, red edges. 7s. 6d.

GREEK AND ENGLISH TESTAMENT. THE STU-

DENT'S EDITION of the above, on *large writing paper.* 4to. 12s.

GREEK TESTAMENT, ex editione Stephani tertia, 1550.

Small 8vo. 3s. 6d.

THE NEW TESTAMENT IN GREEK according to the

text followed in the Authorised Version, with the Variations adopted in the Revised Version. Edited by F. H. A. SCRIVENER M.A., D.C.L., LL.D. Crown 8vo. 6s. Morocco boards or limp. 12s.

THE PARALLEL NEW TESTAMENT GREEK AND

ENGLISH, being the Authorised Version set forth in 1611 Arranged in Parallel Columns with the Revised Version of 1881, and with the original Greek, as edited by F. H. A. SCRIVENER, M.A., D.C.L., LL.D. Prebendary of Exeter and Vicar of Hendon. Crown 8vo. 12s. 6d. *The Revised Version is the Joint Property of the Universities of Cambridge and Oxford.*

London : Cambridge University Press Warehouse, 17 *Paternoster Row.*

THE GOSPEL ACCORDING TO ST MATTHEW in
Anglo-Saxon and Northumbrian Versions, synoptically arranged: with Collations of the best Manuscripts. By J. M. KEMBLE, M.A. and Archdeacon HARDWICK. Demy 4to. 10s.

> NEW EDITION. By the Rev. Professor SKEAT.   [*In the Press.*

THE GOSPEL ACCORDING TO ST MARK in Anglo-
Saxon and Northumbrian Versions, synoptically arranged: with Collations exhibiting all the Readings of all the MSS. Edited by the Rev. Professor SKEAT, M.A. late Fellow of Christ's College, and author of a MŒSO-GOTHIC Dictionary. Demy 4to. 10s.

THE GOSPEL ACCORDING TO ST LUKE, uniform
with the preceding, by the same Editor. Demy 4to. 10s.

THE GOSPEL ACCORDING TO ST JOHN, uniform
with the preceding, by the same Editor. Demy 4to. 10s.

"*The Gospel according to St John, in Anglo-Saxon and Northumbrian Versions:* Edited for the Syndics of the University Press, by the Rev. Walter W. Skeat, M.A., Elrington and Bosworth Professor of Anglo-Saxon in the University of Cambridge, completes an undertaking designed and commenced by that distinguished scholar, J. M. Kemble, some forty years ago. Of the particular volume now before us, we can only say it is worthy of its two predecessors. We repeat that the service rendered to the study of Anglo-Saxon by this Synoptic collection cannot easily be overstated."—*Contemporary Review.*

THE POINTED PRAYER BOOK, being the Book of
Common Prayer with the Psalter or Psalms of David, pointed as they are to be sung or said in Churches. Royal 24mo. Cloth. 1s. 6d.

The same in square 32mo. cloth. 6d.

"The 'Pointed Prayer Book' deserves mention for the new and ingenious system on which the pointing has been marked, and still more for the terseness and clearness of the directions given for using it."—*Times.*

THE CAMBRIDGE PSALTER, for the use of Choirs and
Organists. Specially adapted for Congregations in which the "Cambridge Pointed Prayer Book" is used. Demy 8vo. cloth extra, 3s. 6d. Cloth limp, cut flush. 2s. 6d.

THE PARAGRAPH PSALTER, arranged for the use of
Choirs by BROOKE FOSS WESTCOTT, D.D., Regius Professor of Divinity in the University of Cambridge. Fcap. 4to. 5s.

The same in royal 32mo. Cloth 1s. Leather 1s. 6d.

"The Paragraph Psalter exhibits all the care, thought, and learning that those acquainted with the works of the Regius Professor of Divinity at Cambridge would expect to find, and there is not a clergyman or organist in England who should be without this Psalter as a work of reference."—*Morning Post.*

THE MISSING FRAGMENT OF THE LATIN TRANS-
LATION OF THE FOURTH BOOK OF EZRA, discovered, and edited with an Introduction and Notes, and a facsimile of the MS., by ROBERT L. BENSLY, M.A., Reader in Hebrew, Gonville and Caius College, Cambridge. Demy 4to. 10s.

"Edited with true scholarly completeness."—*Westminster Review.*
"It has been said of this book that it has added a new chapter to the Bible, and, startling as the statement may at first sight appear, it is no exaggeration of the actua fact, if by the Bible we understand that of the larger size which contains the Apocrypha, and if the Second Book of Esdras can be fairly called a part of the Apocrypha."—*Saturday Review.*

THEOLOGY—(ANCIENT).

THE GREEK LITURGIES. Chiefly from original Authorities. By C. A. SWAINSON, D.D., Master of Christ's College, Cambridge. Crown 4to. Paper covers. 15*s.*

THE PALESTINIAN MISHNA. By W. H. LOWE, M.A. Lecturer in Hebrew at Christ's College, Cambridge. Royal 8vo. 21*s.*

SAYINGS OF THE JEWISH FATHERS, comprising Pirqe Aboth and Pereq R. Meir in Hebrew and English, with Critical and Illustrative Notes. By CHARLES TAYLOR, D.D. Master of St John's College, Cambridge, and Honorary Fellow of King's College, London. Demy 8vo. 10*s.*

"The 'Masseketh Aboth' stands at the head of Hebrew non-canonical writings. It is of ancient date, claiming to contain the dicta of teachers who flourished from B.C. 200 to the same year of our era. The precise time of its compilation in its present form is, of course, in doubt. Mr Taylor's explanatory and illustrative commentary is very full and satisfactory." —*Spectator.*
"If we mistake not, this is the first precise translation into the English language, accompanied by scholarly notes, of any portion of the Talmud. In other words, it is the first instance of that most valuable and neglected portion of

Jewish literature being treated in the same way as a Greek classic in an ordinary critical edition... The *Sayings of the Jewish Fathers* may claim to be scholarly, and, moreover, of a scholarship unusually thorough and finished.' —*Dublin University Magazine.*
" A careful and thorough edition which does credit to English scholarship, containing a series of sentences or maxims ascribed mostly to Jewish teachers immediately preceding, or immediately following the Christian era..."—*Contemporary Review.*

THEODORE OF MOPSUESTIA'S COMMENTARY ON THE MINOR EPISTLES OF S. PAUL. The Latin Version with the Greek Fragments, edited from the MSS. with Notes and an Introduction, by H. B. SWETE, D.D., Rector of Ashdon, Essex, and late Fellow of Gonville and Caius College, Cambridge. In Two Volumes. Vol. I., containing the Introduction, with Facsimiles of the MSS., and the Commentary upon Galatians—Colossians. Demy 8vo. 12*s.*

"In dem oben verzeichneten Buche liegt uns die erste Hälfte einer vollständigen, ebenso sorgfältig gearbeiteten wie schön ausgestatteten Ausgabe des Commentars mit ausführlichen Prolegomena und reichhaltigen kritischen und erläuternden Anmerkungen vor."—*Literarisches Centralblatt.*
"It is the result of thorough, careful, and patient investigation of all the points bearing on the subject, and the results are presented with admirable good sense and modesty."—*Guardian.*
"Auf Grund dieser Quellen ist der Text bei Swete mit musterhafter Akribie hergestellt. Aber auch sonst hat der Herausgeber mit unermüdlichem Fleisse und eingehendster Sachkenntniss sein Werk mit allen denjenigen Zugaben ausgerüstet, welche bei einer solchen Text-Ausgabe nur irgend erwartet werden können. . . . Von den drei Haupt-

handschriften . . . sind vortreffliche photographische Facsimile's beigegeben, wie überhaupt das ganze Werk von der *University Press* zu Cambridge mit bekannter Eleganz ausgestattet ist."—*Theologische Literaturzeitung.*
"It is a hopeful sign, amid forebodings which arise about the theological learning of the Universities, that we have before us the first instalment of a thoroughly scientific and painstaking work, commenced at Cambridge and completed at a country rectory."-*Church Quarterly Review* (Jan. 1881).
"Herrn Swete's Leistung ist eine so tüchtige dass wir das Werk in keinen besseren Händen wissen möchten, und mit den sichersten Erwartungen auf das Gelingen der Fortsetzung entgegen sehen."—*Göttingische gelehrte Anzeigen* (Sept. 1881).

VOLUME II., containing the Commentary on 1 Thessalonians—Philemon, Appendices and Indices. 12*s.*

"Eine Ausgabe . . . für welche alle zugänglichen Hülfsmittel in musterhafter Weise benützt wurden . . . eine reife Frucht siebenjährigen Fleisses."—*Theologische Literaturzeitung* (Sept. 23, 1882).

"Mit deiselben Sorgfalt bearbeitet die wir bei dem ersten Theile gerühmt haben."—*Literarisches Centralblatt* (July 29, 1882).

SANCTI IRENÆI EPISCOPI LUGDUNENSIS libros quinque adversus Hæreses, versione Latina cum Codicibus Claromontano ac Arundeliano denuo collata, præmissa de placitis Gnosticorum prolusione, fragmenta necnon Græce, Syriace, Armeniace, commentatione perpetua et indicibus variis edidit W. WIGAN HARVEY, S.T.B. Collegii Regalis olim Socius. 2 Vols. Demy 8vo. 18s.

M. MINUCII FELICIS OCTAVIUS. The text newly revised from the original MS., with an English Commentary, Analysis, Introduction, and Copious Indices. Edited by H. A. HOLDEN, LL.D. late Head Master of Ipswich School, formerly Fellow of Trinity College, Cambridge. Crown 8vo. 7s. 6d.

THEOPHILI EPISCOPI ANTIOCHENSIS LIBRI TRES AD AUTOLYCUM edidit, Prolegomenis Versione Notulis Indicibus instruxit GULIELMUS GILSON HUMPHRY, S.T.B. Collegii Sanctiss. Trin. apud Cantabrigienses quondam Socius. Post 8vo. 5s.

THEOPHYLACTI IN EVANGELIUM S. MATTHÆI COMMENTARIUS, edited by W. G. HUMPHRY, B.D. Prebendary of St Paul's, late Fellow of Trinity College. Demy 8vo. 7s. 6d.

TERTULLIANUS DE CORONA MILITIS, DE SPECTACULIS, DE IDOLOLATRIA, with Analysis and English Notes, by GEORGE CURREY, D.D. Preacher at the Charter House, late Fellow and Tutor of St John's College. Crown 8vo. 5s.

THEOLOGY—(ENGLISH).

WORKS OF ISAAC BARROW, compared with the Original MSS., enlarged with Materials hitherto unpublished. A new Edition, by A. NAPIER, M.A. of Trinity College, Vicar of Holkham, Norfolk. 9 Vols. Demy 8vo. £3. 3s.

TREATISE OF THE POPE'S SUPREMACY, and a Discourse concerning the Unity of the Church, by ISAAC BARROW. Demy 8vo. 7s. 6d.

PEARSON'S EXPOSITION OF THE CREED, edited by TEMPLE CHEVALLIER, B.D. late Fellow and Tutor of St Catharine's College, Cambridge. New Edition. Revised by R. Sinker, B.D., Librarian of Trinity College. Demy 8vo. 12s.

"A new edition of Bishop Pearson's famous work *On the Creed* has just been issued by the Cambridge University Press. It is the well-known edition of Temple Chevallier, thoroughly overhauled by the Rev. R. Sinker, of Trinity College. The whole text and notes have been most carefully examined and corrected, and special pains have been taken to verify the almost innumerable references. These have been more clearly and accurately given in very many places, and the citations themselves have been adapted to the best and newest texts of the several authors—·texts which have undergone vast improvements within the last two centuries. The Indices have also been revised and enlarged......Altogether this appears to be the most complete and convenient edition as yet published of a work which has long been recognised in all quarters as a standard one."— *Guardian.*

AN ANALYSIS OF THE EXPOSITION OF THE

CREED written by the Right Rev. JOHN PEARSON, D.D. late Lord Bishop of Chester, by W. H. MILL, D.D. late Regius Professor of Hebrew in the University of Cambridge. Demy 8vo. 5s.

WHEATLY ON THE COMMON PRAYER, edited by

G. E. CORRIE, D.D. Master of Jesus College, Examining Chaplain to the late Lord Bishop of Ely. Demy 8vo. 7s. 6d.

CÆSAR MORGAN'S INVESTIGATION OF THE

TRINITY OF PLATO, and of Philo Judæus, and of the effects which an attachment to their writings had upon the principles and reasonings of the Fathers of the Christian Church. Revised by H. A. HOLDEN, LL.D., formerly Fellow of Trinity College, Cambridge. Crown 8vo. 4s.

TWO FORMS OF PRAYER OF THE TIME OF QUEEN

ELIZABETH. Now First Reprinted. Demy 8vo. 6d.

"From 'Collections and Notes' 1867—1876, by W. Carew Hazlitt (p. 340), we learn that— 'A very remarkable volume, in the original vellum cover, and containing 25 Forms of Prayer of the reign of Elizabeth, each with the autograph of Humphrey Dyson, has lately fallen into the hands of my friend Mr H. Pyne. It is mentioned specially in the Preface to the Par- ker Society's volume of Occasional Forms of Prayer, but it had been lost sight of for 200 years.' By the kindness of the present pos- sessor of this valuable volume, containing in all 25 distinct publications, I am enabled to re- print in the following pages the two Forms of Prayer supposed to have been lost."—*Ex- tract from the* PREFACE.

SELECT DISCOURSES, by JOHN SMITH, late Fellow of

Queens' College, Cambridge. Edited by H. G. WILLIAMS, B.D. late Professor of Arabic. Royal 8vo. 7s. 6d.

"The 'Select Discourses' of John Smith, collected and published from his papers after his death, are, in my opinion, much the most considerable work left to us by this Cambridge School [the Cambridge Platonists]. They have a right to a place in English literary history."—Mr MATTHEW ARNOLD, in the *Contempo- rary Review.*
"Of all the products of the Cambridge School, the 'Select Discourses' are perhaps the highest, as they are the most accessible and the most widely appreciated...and indeed no spiritually thoughtful mind can read them unmoved. They carry us so directly into an atmosphere of divine philosophy, luminous with the richest lights of meditative genius... He was one of those rare thinkers in whom largeness of view, and depth, and wealth of poetic and speculative insight, only served to evoke more fully the religious spirit, and while he drew the mould of his thought from Plotinus, he vivified the substance of it from St Paul."—Principal TULLOCH, *Rational Theology in England in the 17th Century.*
"We may instance Mr Henry Griffin Wil- liams's revised edition of Mr John Smith's 'Select Discourses,' which have won Mr Matthew Arnold's admiration, as an example of worthy work for an University Press to undertake."—*Times.*

THE HOMILIES, with Various Readings, and the Quo-

tations from the Fathers given at length in the Original Languages. Edited by G. E. CORRIE, D.D., Master of Jesus College. Demy 8vo. 7s. 6d.

DE OBLIGATIONE CONSCIENTIÆ PRÆLECTIONES

decem Oxonii in Schola Theologica habitæ a ROBERTO SANDERSON, SS. Theologiæ ibidem Professore Regio. With English Notes, in- cluding an abridged Translation, by W. WHEWELL, D.D. late Master of Trinity College. Demy 8vo. 7s. 6d.

ARCHBISHOP USHER'S ANSWER TO A JESUIT,

with other Tracts on Popery. Edited by J. SCHOLEFIELD, M.A. late Regius Professor of Greek in the University. Demy 8vo. 7s. 6d.

WILSON'S ILLUSTRATION OF THE METHOD OF

explaining the New Testament, by the early opinions of Jews and Christians concerning Christ. Edited by T. TURTON, D.D. late Lord Bishop of Ely. Demy 8vo. 5s.

LECTURES ON DIVINITY delivered in the University

of Cambridge, by JOHN HEY, D.D. Third Edition, revised by T. TURTON, D.D. late Lord Bishop of Ely. 2 vols. Demy 8vo. 15s.

ARABIC, SANSKRIT AND SYRIAC.

POEMS OF BEHÁ ED DÍN ZOHEIR OF EGYPT.

With a Metrical Translation, Notes and Introduction, by E. H. PALMER, M.A., Barrister-at-Law of the Middle Temple, late Lord Almoner's Professor of Arabic, formerly Fellow of St John's College, Cambridge. 3 vols. Crown 4to.

    Vol. I. The ARABIC TEXT. 10s. 6d.; Cloth extra. 15s.
    Vol. II. ENGLISH TRANSLATION. 10s. 6d.; Cloth extra. 15s.

"We have no hesitation in saying that in both Prof. Palmer has made an addition to Oriental literature for which scholars should be grateful; and that, while his knowledge of Arabic is a sufficient guarantee for his mastery of the original, his English compositions are distinguished by versatility, command of language, rhythmical cadence, and, as we have remarked, by not unskilful imitations of the styles of several of our own favourite poets, living and dead."—*Saturday Review.*

"This sumptuous edition of the poems of Behá-ed-dín Zoheir is a very welcome addition to the small series of Eastern poets accessible to readers who are not Orientalists."—*Academy.*

KALÍLAH AND DIMNAH, OR, THE FABLES OF

PILPAI; being an account of their literary history, together with an English Translation of the same, with Notes, by I. G. N. KEITH-FALCONER, M.A., Trinity College, formerly Tyrwhitt's Hebrew Scholar. Demy 8vo. *[In the Press.*

THE CHRONICLE OF JOSHUA THE STYLITE, com-

posed in Syriac A.D. 507 with an English translation and notes, by W. WRIGHT, LL.D., Professor of Arabic. Demy 8vo. 10s. 6d.

"Die lehrreiche kleine Chronik Josuas hat nach Assemani und Martin in Wright einen dritten Bearbeiter gefunden, der sich um die Emendation des Textes wie um die Erklärung der Realien wesentlich verdient gemacht hat ... Ws. Josua-Ausgabe ist eine sehr dankenswerte Gabe und besonders empfehlenswert als ein Lehrmittel für den syrischen Unterricht; es erscheint auch gerade zur rechten Zeit, da die zweite Ausgabe von Roedigers syrischer Chrestomathie im Buchhandel vollständig vergriffen und diejenige von Kirsch-Bernstein nur noch in wenigen Exemplaren vorhanden ist."—*Deutsche Litteraturzeitung.*

NALOPÁKHYÁNAM, OR, THE TALE OF NALA;

containing the Sanskrit Text in Roman Characters, followed by a Vocabulary in which each word is placed under its root, with references to derived words in Cognate Languages, and a sketch of Sanskrit Grammar. By the late Rev. THOMAS JARRETT, M.A. Trinity College, Regius Professor of Hebrew. Demy 8vo. 10s.

NOTES ON THE TALE OF NALA, for the use of

Classical Students, by J. PEILE, M.A. Fellow and Tutor of Christ's College. Demy 8vo. 12s.

CATALOGUE OF THE BUDDHIST SANSKRIT

MANUSCRIPTS in the University Library, Cambridge. Edited by C. BENDALL, M.A., Fellow of Gonville and Caius College. Demy 8vo. 12s.

"It is unnecessary to state how the compilation of the present catalogue came to be placed in Mr Bendall's hands; from the character of his work it is evident the selection was judicious, and we may fairly congratulate those concerned in it on the result ... Mr Bendall has entitled himself to the thanks of all Oriental scholars, and we hope he may have before him a long course of successful labour in the field he has chosen."—*Athenæum.*

GREEK AND LATIN CLASSICS, &c. (See also pp. 24—27.)

SOPHOCLES: The Plays and Fragments, with Critical Notes, Commentary, and Translation in English Prose, by R. C. JEBB, M.A., LL.D., Professor of Greek in the University of Glasgow.

Part I. Oedipus Tyrannus. Demy 8vo. 15s.

"In undertaking, therefore, to interpret Sophocles to the classical scholar and to the British public, Professor Jebb expounds the most consummate poetical artist of what common consent allows to be the highest stage in Greek culture... As already hinted, Mr Jebb in his work aims at two classes of readers. He keeps in view the Greek student and the English scholar who knows little or no Greek. His critical notes and commentary are meant for the first class . . . The present edition of Sophocles is to consist of eight volumes—one being allowed to each play—and the eighth, containing the fragments and a series of short essays on subjects of general interest relating to Sophocles. If the remaining volumes maintain the high level of the present one, it will, when completed, be truly an *edition de luxe.*"—*Glasgow Herald.*

AESCHYLI FABULAE.—ΙΚΕΤΙΔΕΣ ΧΟΗΦΟΡΟΙ IN LIBRO MEDICEO MENDOSE SCRIPTAE EX VV. DD. CONIECTURIS EMENDATIUS EDITAE cum Scholiis Graecis et brevi adnotatione critica, curante F. A. PALEY, M.A., LL.D. Demy 8vo. 7s. 6d.

THE AGAMEMNON OF AESCHYLUS. With a Translation in English Rhythm, and Notes Critical and Explanatory. **New Edition Revised.** By BENJAMIN HALL KENNEDY, D.D., Regius Professor of Greek. Crown 8vo. 6s.

"One of the best editions of the masterpiece of Greek tragedy."—*Athenæum.*
"It is needless to multiply proofs of the value of this volume alike to the poetical translator, the critical scholar, and the ethical student."—*Saturday Review.*

THE THEÆTETUS OF PLATO with a Translation and Notes by the same Editor. Crown 8vo. 7s. 6d.

PLATO'S PHÆDO, literally translated, by the late E. M. COPE, Fellow of Trinity College, Cambridge. Demy 8vo. 5s.

ARISTOTLE.—ΠΕΡΙ ΔΙΚΑΙΟΣΤΝΗΣ. THE FIFTH BOOK OF THE NICOMACHEAN ETHICS OF ARISTOTLE. Edited by HENRY JACKSON, M.A., Fellow of Trinity College, Cambridge. Demy 8vo. 6s.

"It is not too much to say that some of the points he discusses have never had so much light thrown upon them before. . . . Scholars will hope that this is not the only portion of the Aristotelian writings which he is likely to edit."—*Athenæum.*

ARISTOTLE.—ΠΕΡΙ ΨΤΧΗΣ. ARISTOTLE'S PSYCHOLOGY, in Greek and English, with Introduction and Notes, by EDWIN WALLACE, M.A., Fellow and Tutor of Worcester College, Oxford. Demy 8vo. 18s.

"In an elaborate introduction Mr Wallace collects and correlates all the passages from the various works of Aristotle bearing on these points, and this he does with a width of learning that marks him out as one of our foremost Aristotlic scholars, and with a critical acumen that is far from common."—*Glasgow Herald.*
"As a clear exposition of the opinions of Aristotle on psychology, Mr Wallace's work is of distinct value—the introduction is excellently wrought out, the translation is good, the notes are thoughtful, scholarly, and full. We therefore can welcome a volume like this, which is useful both to those who study it as scholars, and to those who read it as students of philosophy."—*Scotsman.*
"The notes are exactly what such notes ought to be,—helps to the student, not mere displays of learning. By far the more valuable parts of the notes are neither critical nor literary, but philosophical and expository of the thought, and of the connection of thought, in the treatise itself. In this relation the notes are invaluable. Of the translation, it may be said that an English reader may fairly master by means of it this great treatise of Aristotle."—*Spectator.*

A SELECTION OF GREEK INSCRIPTIONS, with Introductions and Annotations by E. S. ROBERTS, M.A. Fellow and Tutor of Gonville and Caius College. [*In the Press.*

PINDAR. OLYMPIAN AND PYTHIAN ODES. With Notes Explanatory and Critical, Introductions and Introductory Essays. Edited by C. A. M. FENNELL, M.A., late Fellow of Jesus College. Crown 8vo. 9s.

"Mr Fennell deserves the thanks of all classical students for his careful and scholarly edition of the Olympian and Pythian odes. He brings to his task the necessary enthusiasm for his author, great industry, a sound judgment, and, in particular, copious and minute learning in comparative philology. To his qualifications in this last respect every page bears witness."—*Athenæum.*

"Considered simply as a contribution to the study and criticism of Pindar, Mr Fennell's edition is a work of great merit... Altogether, this edition is a welcome and wholesome sign

of the vitality and development of Cambridge scholarship, and we are glad to see that it is to be continued."—*Saturday Review.*

"Mr C. A. M. Fennell's 'Pindar' displays that union of laborious research and unassuming directness of style which characterizes the best modern scholarship... The notes, which are in English, and at the foot of each page, are clear and to the point. There is an introduction to each Ode. There are Greek and English Indices, and an Index of Quotations."—*Westminster Review.*

—— THE ISTHMIAN AND NEMEAN ODES. By the same Editor. Crown 8vo. 9s.

"Encouraged by the warm praise with which Mr Fennell's edition of the Olympian and Pythian odes was everywhere received, the Pitt Press Syndicate very properly invited him to continue his work and edit the remainder of Pindar... His notes are full of original ideas carefully worked out, and if he often adopts the opinion of other editors, he does not do so without making it sufficiently plain that he has discussed the question for himself and decided it upon the evidence. As a handy and instructive edition of a difficult classic no work of recent years surpasses Mr

Fennell's 'Pindar.'"—*Athenæum.*

"Mr Fennell, whose excellent edition of the Olympian and Pythian Odes of Pindar appeared some four years ago, has now published the Nemean and Isthmian Odes, together with a selection from the extant fragments of Pindar. This work is in no way inferior to the previous volume. The commentary affords valuable help to the study of the most difficult of Greek authors, and is enriched with notes on points of scholarship and etymology which could only have been written by a scholar of very high attainments."—*Saturday Review.*

ARISTOTLE. THE RHETORIC. With a Commentary by the late E. M. COPE, Fellow of Trinity College, Cambridge, revised and edited by J. E. SANDYS, M.A., Fellow and Tutor of St John's College, Cambridge, and Public Orator. With a biographical Memoir by H. A. J. MUNRO, M.A. Three Volumes, Demy 8vo. £1. 11s. 6d.

"This work is in many ways creditable to the University of Cambridge. If an English student wishes to have a full conception of what is contained in the *Rhetoric* of Aristotle, to Mr Cope's edition he must go."—*Academy.*

"Mr Sandys has performed his arduous duties with marked ability and admirable tact. ... In every part of his work—revising, supplementing, and completing—he has done exceedingly well."—*Examiner.*

PRIVATE ORATIONS OF DEMOSTHENES, with Introductions and English Notes, by F. A. PALEY, M.A. Editor of Aeschylus, etc. and J. E. SANDYS, M.A. Fellow and Tutor of St John's College, and Public Orator in the University of Cambridge.

PART I. Contra Phormionem, Lacritum, Pantaenetum, Boeotum de Nomine, Boeotum de Dote, Dionysodorum. Crown 8vo. 6s.

PART II. Pro Phormione, Contra Stephanum I. II.; Nicostratum, Cononem, Calliclem. Crown 8vo. 7s. 6d.

DEMOSTHENES AGAINST ANDROTION AND AGAINST TIMOCRATES, with Introductions and English Commentary, by WILLIAM WAYTE, M.A., late Professor of Greek, University College, London, Formerly Fellow of King's College, Cambridge, and Assistant Master at Eton. Crown 8vo. 7s. 6d.

"The present edition may therefore be used by students more advanced than school-boys, and to their purposes it is admirably suited. There is an excellent introduction to and analysis of each speech, and at the beginning of

each paragraph of the text there is a summary of its subject-matter... The notes are uniformly good, whether they deal with questions of scholarship or with points of Athenian law."—*Saturday Review.*

THE TYPES OF GREEK COINS. By PERCY GARDNER,
M.A., F.S.A., Disney Professor of Archæology. With 16 Autotype plates, containing photographs of Coins of all parts of the Greek World. Impl. 4to. Cloth extra, £1. 11s. 6d.; Roxburgh (Morocco back), £2. 2s.

"Professor Gardner's book is written with such lucidity and in a manner so straightforward that it may well win converts, and it may be distinctly recommended to that omnivorous class of readers—'men in the schools.' The history of ancient coins is so interwoven with and so vividly illustrates the history of ancient States, that students of Thucydides and Herodotus cannot afford to neglect Professor Gardner's introduction to Hellenic numismatics . . . The later part of Mr Gardner's useful and interesting volume is devoted to the artistic and archæological aspect of coins, and can scarcely be studied apart from photographs (like those which he supplies) or casts of the original medals."—*Saturday Review.*

'The Types of Greek Coins' is a work which is less purely and dryly scientific. Nevertheless, it takes high rank as proceeding upon a truly scientific basis at the same time that it treats the subject of numismatics in an attractive style and is elegant enough to justify its appearance in the drawing-room . . . . Sixteen autotype plates reproduce with marvellous reality more than six hundred types of picked specimens of coins in every style, from the cabinets of the British Museum and other collections.—*Athenæum.*

THE BACCHAE OF EURIPIDES. With Introduction,
Critical Notes, and Archæological Illustrations, by J. E. SANDYS, M.A., Fellow and Tutor of St John's College, Cambridge, and Public Orator. Crown 8vo. 10s. 6d.

"Of the present edition of the *Bacchæ* by Mr Sandys we may safely say that never before has a Greek play, in England at least, had fuller justice done to its criticism, interpretation, and archæological illustration, whether for the young student or the more advanced scholar. The Cambridge Public Orator may be said to have taken the lead in issuing a complete edition of a Greek play, which is destined perhaps to gain redoubled favour now that the study of ancient monuments has been applied to its illustration."—*Saturday Review.*

"The volume is interspersed with well-executed woodcuts, and its general attractiveness of form reflects great credit on the University Press. In the notes Mr Sandys has more than sustained his well-earned reputation as a careful and learned editor, and shows considerable advance in freedom and lightness of style. . . . Under such circumstances it is superfluous to say that for the purposes of teachers and advanced students this handsome edition far surpasses all its predecessors."—*Athenæum.*

"It has not, like so many such books, been hastily produced to meet the momentary need of some particular examination; but it has employed for some years the labour and thought of a highly finished scholar, whose aim seems to have been that his book should go forth *totus teres atque rotundus*, armed at all points with all that may throw light upon its subject. The result is a work which will not only assist the schoolboy or undergraduate in his tasks, but will adorn the library of the scholar."—*The Guardian.*

ESSAYS ON THE ART OF PHEIDIAS. By C. WALD-
STEIN, M.A., Phil. D., Reader in Classical Archæology in the University of Cambridge. Royal 8vo. With Illustrations.

[In the Press.

M. TULLI CICERONIS DE FINIBUS BONORUM
ET MALORUM LIBRI QUINQUE. The text revised and explained; With a Translation by JAMES S. REID, M.L., Fellow and Assistant Tutor of Gonville and Caius College. In three Volumes.

[In the Press.

VOL. III. Containing the Translation. Demy 8vo. 8s.

M. T. CICERONIS DE OFFICIIS LIBRI TRES,
with Marginal Analysis, an English Commentary, and copious Indices, by H. A. HOLDEN, LL.D., late Fellow of Trinity College, Cambridge. **Fifth Edition.** Revised and considerably enlarged. Crown 8vo. 9s.

"Dr Holden has issued an edition of what is perhaps the easiest and most popular of Cicero's philosophical works, the *de Officiis*, which, especially in the form which it has now assumed after two most thorough revisions, leaves little or nothing to be desired in the fullness and accuracy of its treatment alike of the matter and the language."—*Academy.*

M. TVLLI CICERONIS PRO C RABIRIO [PERDVEL-
LIONIS REO] ORATIO AD QVIRITES With Notes Introduction and Appendices by W E HEITLAND MA, Fellow and Lecturer of St John's College, Cambridge. Demy 8vo. 7s. 6d.

M. TULLII CICERONIS DE NATURA DEORUM

Libri Tres, with Introduction and Commentary by JOSEPH B. MAYOR, M.A., late Professor of Moral Philosophy at King's College, London, together with a new collation of several of the English MSS. by J. H. SWAINSON, M.A., formerly Fellow of Trinity College, Cambridge. Vol. I. Demy 8vo. 10s. 6d.   Vol. II. 12s. 6d.

"Such editions as that of which Prof. Mayor has given us the first instalment will doubtless do much to remedy this undeserved neglect. It is one on which great pains and much learning have evidently been expended, and is in every way admirably suited to meet the needs of the student . . . The notes of the editor are all that could be expected from his well-known learning and scholarship."—*Academy*.

P. VERGILI MARONIS OPERA cum Prolegomenis

et Commentario Critico pro Syndicis Preli Academici edidit BENJAMIN HALL KENNEDY, S.T.P., Graecae Linguae Professor Regius. Extra Fcap. 8vo. 5s.

MATHEMATICS, PHYSICAL SCIENCE, &c.

MATHEMATICAL AND PHYSICAL PAPERS. By

Sir W. THOMSON, LL.D., D.C.L., F.R.S., Professor of Natural Philosophy in the University of Glasgow. Collected from different Scientific Periodicals from May 1841, to the present time. Vol. I. Demy 8vo. 18s.   [Vol. II. *In the Press.*

"Wherever exact science has found a follower Sir William Thomson's name is known as a leader and a master. For a space of 40 years each of his successive contributions to knowledge in the domain of experimental and mathematical physics has been recognized as marking a stage in the progress of the subject. But, unhappily for the mere learner, he is no writer of text-books. His eager fertility overflows into the nearest available journal . . . The papers in this volume deal largely with the subject of the dynamics of heat. They begin with two or three articles which were in part written at the age of 17, before the author had commenced residence as an undergraduate in Cambridge . . . No student of mechanical engineering, who aims at the higher levels of his profession, can afford to be ignorant of the principles and methods set forth in these great memoirs . . . The article on the absolute measurement of electric and galvanic quantities (1851) has

borne rich and abundant fruit. Twenty years after its date the International Conference of Electricians at Paris, assisted by the author himself, elaborated and promulgated a series of rules and units which are but the detailed outcome of the principles laid down in these papers."—*The Times*.

"We are convinced that nothing has had a greater effect on the progress of the theories of electricity and magnetism during the last ten years than the publication of Sir W. Thomson's reprint of papers on electrostatics and magnetism, and we believe that the present volume is destined in no less degree to further the advancement of physical science. We owe the modern dynamical theory of heat almost wholly to Joule and Thomson, and Clausius and Rankine, and we have here collected together the whole of Thomson's investigations on this subject, together with the papers published jointly by himself and Joule."—*Glasgow Herald.*

MATHEMATICAL AND PHYSICAL PAPERS, by

GEORGE GABRIEL STOKES, M.A., D.C.L., LL.D., F.R.S., Fellow of Pembroke College, and Lucasian Professor of Mathematics in the University of Cambridge. Reprinted from the Original Journals and Transactions, with Additional Notes by the Author. Vol. I. Demy 8vo. 15s. VOL. II. 15s.

"The volume of Professor Stokes's papers contains much more than his hydrodynamical papers. The undulatory theory of light is treated, and the difficulties connected with its application to certain phenomena, such as aberration, are carefully examined and resolved. Such difficulties are commonly passed over with scant notice in the text-books . . . Those to whom difficulties like these are real stumbling-blocks will still turn for enlightenment to Professor Stokes's old, but still fresh and still

necessary, dissertations. There nothing is slurred over, nothing extenuated. We learn exactly the weaknesses of the theory, and the direction in which the completer theory of the future must be sought for. The same spirit pervades the papers on pure mathematics which are included in the volume. They have a severe accuracy of style which well befits the subtle nature of the subjects, and inspires the completest confidence in their author."—*The Times*.

VOLUME III.   *In the Press.*

THE SCIENTIFIC PAPERS OF THE LATE PROF.

J. CLERK MAXWELL. Edited by W. D. NIVEN, M.A. In 2 vols. Royal 4to.   [*In the Press.*

A TREATISE ON NATURAL PHILOSOPHY. By

Sir W. THOMSON, LL.D., D.C.L., F.R.S., Professor of Natural Philosophy in the University of Glasgow. and P. G. TAIT, M.A., Professor of Natural Philosophy in the University of Edinburgh. Part I. Demy 8vo. 16s.

" In this, the second edition, we notice a large amount of new matter, the importance of which is such that any opinion which we could

form within the time at our disposal would be utterly inadequate."—*Nature.*

Part II. Demy 8vo. 18s.

ELEMENTS OF NATURAL PHILOSOPHY. By Pro-

fessors Sir W. THOMSON and P. G. TAIT. Part I. Demy 8vo. *Second Edition.* 9s.

A TREATISE ON THE THEORY OF DETERMI-

NANTS AND THEIR APPLICATIONS IN ANALYSIS AND GEOMETRY, by ROBERT FORSYTH SCOTT, M.A., of St John's College, Cambridge. Demy 8vo. 12s.

" This able and comprehensive treatise will be welcomed by the student as bringing within his reach the results of many important re-

searches on this subject which have hitherto been for the most part inaccessible to him."— *Athenæum.*

HYDRODYNAMICS, a Treatise on the Mathematical

Theory of the Motion of Fluids, by HORACE LAMB, M.A., formerly Fellow of Trinity College, Cambridge ; Professor of Mathematics in the University of Adelaide. Demy 8vo. 12s.

THE ANALYTICAL THEORY OF HEAT, by JOSEPH

FOURIER. Translated, with Notes, by A. FREEMAN, M.A., Fellow of St John's College, Cambridge. Demy 8vo. 16s.

" It is time that Fourier's masterpiece, *The Analytical Theory of Heat,* translated by Mr Alex. Freeman, should be introduced to those English students of Mathematics who do not follow with freedom a treatise in any language but their own. It is a model of mathematical reasoning applied to physical phenomena, and is remarkable for the ingenuity of the analytical

process employed by the author."—*Contemporary Review,* October, 1878.

" There cannot be two opinions as to the value and importance of the *Théorie de la Chaleur* . . . It is still *the* text-book of Heat Conduction, and there seems little present prospect of its being superseded, though it is already more than half a century old."—*Nature.*

THE ELECTRICAL RESEARCHES OF THE Honour-

able HENRY CAVENDISH, F.R.S. Written between 1771 and 1781. Edited from the original manuscripts in the possession of the Duke of Devonshire, K. G., by the late J. CLERK MAXWELL, F.R.S. Demy 8vo. 18s.

" Every department of editorial duty appears to have been most conscientiously performed ; and it must have been no small satis-

faction to Prof. Maxwell to see this goodly volume completed before his life's work was done."—*Athenæum.*

AN ELEMENTARY TREATISE ON QUATERNIONS.

By P. G. TAIT, M.A., Professor of Natural Philosophy in the University of Edinburgh. *Second Edition.* Demy 8vo. 14s.

THE MATHEMATICAL WORKS OF ISAAC BAR-

ROW, D.D. Edited by W. WHEWELL, D.D. Demy 8vo. 7s. 6d.

AN ATTEMPT TO TEST THE THEORIES OF

CAPILLARY ACTION by FRANCIS BASHFORTH, B.D., late Professor of Applied Mathematics to the Advanced Class of Royal Artillery Officers, Woolwich, and J. C. ADAMS, M.A., F.R.S. Demy 4to. £1. 1s.

NOTES ON QUALITATIVE ANALYSIS. Concise and

Explanatory. By H. J. H. FENTON, M.A., F.I.C., F.C.S., Demonstrator of Chemistry in the University of Cambridge. Late Scholar of Christ's College. Crown 4to. 7s. 6d.

A TREATISE ON THE GENERAL PRINCIPLES OF
CHEMISTRY, by M. M. PATTISON MUIR, M.A., Fellow and Præ-
lector in Chemistry of Gonville and Caius College. Demy 8vo.
[In the Press.

A TREATISE ON THE PHYSIOLOGY OF PLANTS,
by S. H. VINES, M.A., Fellow of Christ's College. *[In the Press.*

THE FOSSILS AND PALÆONTOLOGICAL AFFIN-
ITIES OF THE NEOCOMIAN DEPOSITS OF UPWARE
AND BRICKHILL with Plates, being the Sedgwick Prize Essay
for the Year 1879. By WALTER KEEPING, M.A., F.G.S. Demy 8vo.
10s. 6d.

COUNTERPOINT. A Practical Course of Study, by Professor Sir G. A. MACFARREN, M.A., Mus. Doc. Fourth Edition,
revised. Demy 4to. 7s. 6d.

ASTRONOMICAL OBSERVATIONS made at the Obser-
vatory of Cambridge by the Rev. JAMES CHALLIS, M.A., F.R.S.,
F.R.A.S., late Plumian Professor of Astronomy and Experimental
Philosophy in the University of Cambridge. For various Years, from
1846 to 1860.

ASTRONOMICAL OBSERVATIONS from 1861 to 1865.
Vol. XXI. Royal 4to. 15s. From 1866 to 1869. Vol. XXII.
Royal 4to. *[Nearly ready.*

A CATALOGUE OF THE COLLECTION OF BIRDS
formed by the late H. E. STRICKLAND, now in the possession of the
University of Cambridge. By OSBERT SALVIN, M.A., F.R.S., &c.
Strickland Curator in the University of Cambridge. Demy 8vo. £1. 1s.

"The discriminating notes which Mr Salvin
has here and there introduced make the book
indispensable to every worker on what the
Americans call "the higher plane" of the
science of birds."—*Academy.*

"The author has formed a definite and, as
it seems to us, a righteous idea of what the
catalogue of a collection should be, and, allow-
ing for some occasional slips, has effectively
carried it out."—*Notes and Queries.*

A CATALOGUE OF AUSTRALIAN FOSSILS (in-
cluding Tasmania and the Island of Timor), Stratigraphically and
Zoologically arranged, by R. ETHERIDGE, Jun., F.G.S., Acting Palæ-
ontologist, H.M. Geol. Survey of Scotland. Demy 8vo. 10s. 6d.

"The work is arranged with great clearness,
and contains a full list of the books and papers

consulted by the author, and an index to the
genera."—*Saturday Review.*

ILLUSTRATIONS OF COMPARATIVE ANATOMY,
VERTEBRATE AND INVERTEBRATE, for the Use of Stu-
dents in the Museum of Zoology and Comparative Anatomy. Second
Edition. Demy 8vo. 2s. 6d.

A SYNOPSIS OF THE CLASSIFICATION OF THE
BRITISH PALÆOZOIC ROCKS, by the Rev. ADAM SEDGWICK,
M.A., F.R.S., and FREDERICK McCOY, F.G.S. One vol., Royal 4to.
Plates, £1. 1s.

A CATALOGUE OF THE COLLECTION OF CAM-
BRIAN AND SILURIAN FOSSILS contained in the Geological
Museum of the University of Cambridge, by J. W. SALTER, F.G.S.
With a Portrait of PROFESSOR SEDGWICK. Royal 4to. 7s. 6d.

CATALOGUE OF OSTEOLOGICAL SPECIMENS con-
tained in the Anatomical Museum of the University of Cambridge.
Demy 8vo. 2s. 6d.

LAW.

AN ANALYSIS OF CRIMINAL LIABILITY. By E. C.
CLARK, LL.D., Regius Professor of Civil Law in the University of Cambridge, also of Lincoln's Inn, Barrister-at-Law. Crown 8vo. 7s. 6d.

" Prof. Clark's little book is the substance of lectures delivered by him upon those portions of Austin's work on jurisprudence which deal with the "operation of sanctions"...

Students of jurisprudence will find much to interest and instruct them in the work of Prof. Clark."—*Athenæum.*

PRACTICAL JURISPRUDENCE, a Comment on AUSTIN.
By E. C. CLARK, LL.D. Regius Professor of Civil Law. Crown 8vo. 9s.

A SELECTION OF THE STATE TRIALS. By J. W.
WILLIS-BUND, M.A., LL.B., Barrister-at-Law, Professor of Constitutional Law and History, University College, London. Vol. I. Trials for Treason (1327—1660). Crown 8vo. 18s.

"Mr Willis-Bund has edited 'A Selection of Cases from the State Trials' which is likely to form a very valuable addition to the standard literature... There can be no doubt, therefore, of the interest that can be found in the State trials. But they are large and unwieldy, and it is impossible for the general reader to come across them. Mr Willis-Bund has therefore done good service in making a selection that is in the first volume reduced to a commodious form."—*The Examiner.*
"This work is a very useful contribution to that important branch of the constitutional history of England which is concerned with the growth and development of the law of treason,

as it may be gathered from trials before the ordinary courts. The author has very wisely distinguished these cases from those of impeachment for treason before Parliament, which he proposes to treat in a future volume under the general head 'Proceedings in Parliament.'"
— *The Academy.*
"This is a work of such obvious utility that the only wonder is that no one should have undertaken it before ... In many respects therefore, although the trials are more or less abridged, this is for the ordinary student's purpose not only a more handy, but a more useful work than Howell's."—*Saturday Review.*

VOL. II. In two parts. Price 14s. each.

. "But, although the book is most interesting to the historian of constitutional law, it is also not without considerable value to those who seek information with regard to procedure and the growth of the law of evidence. We should add that Mr Willis-Bund has given short prefaces and appendices to the trials, so as to form a connected narrative of the events in history to which they relate. We can thoroughly recommend the book."—*Law Times.*
"To a large class of readers Mr Willis-Bund's compilation will thus be of great assistance, for he presents in a convenient form a

judicious selection of the principal statutes and the leading cases bearing on the crime of treason ... For all classes of readers these volumes possess an indirect interest, arising from the nature of the cases themselves, from the men who were actors in them, and from the numerous points of social life which are incidentally illustrated in the course of the trials. On these features we have not dwelt, but have preferred to show that the book is a valuable contribution to the study of the subject with which it professes to deal, namely, the history of the law of treason."—*Athenæum.*

Vol. III. *In the Press.*

THE FRAGMENTS OF THE PERPETUAL EDICT
OF SALVIUS JULIANUS, collected, arranged, and annotated by BRYAN WALKER, M.A., LL.D., Law Lecturer of St John's College, and late Fellow of Corpus Christi College, Cambridge. Crown 8vo. 6s.

"In the present book we have the fruits of the same kind of thorough and well-ordered study which was brought to bear upon the notes to the Commentaries and the Institutes . . . Hitherto the Edict has been almost inaccessible to the ordinary English student, and

such a student will be interested as well as perhaps surprised to find how abundantly the extant fragments illustrate and clear up points which have attracted his attention in the Commentaries, or the Institutes, or the Digest."—*Law Times.*

AN INTRODUCTION TO THE STUDY OF JUS-
TINIAN'S DIGEST. Containing an account of its composition
and of the Jurists used or referred to therein, together with a full
Commentary on one Title (de usufructu), by HENRY JOHN ROBY,
M.A., formerly Classical Lecturer in St John's College, Cambridge,
and Professor of Jurisprudence in University College, London.

[In the Press.

THE COMMENTARIES OF GAIUS AND RULES OF
ULPIAN. (**New Edition, revised and enlarged.**) With a Trans-
lation and Notes, by J. T. ABDY, LL.D., Judge of County Courts,
late Regius Professor of Laws in the University of Cambridge, and
BRYAN WALKER, M.A., LL.D., Law Lecturer of St John's College,
Cambridge, formerly Law Student of Trinity Hall and Chancellor's
Medallist for Legal Studies. Crown 8vo. 16*s.*

"As scholars and as editors Messrs Abdy and Walker have done their work well . . . For one thing the editors deserve special commendation. They have presented Gaius to the reader with few notes and those merely by way of reference or necessary explanation. Thus the Roman jurist is allowed to speak for himself, and the reader feels that he is really studying Roman law in the original, and not a fanciful representation of it."—*Athenæum.*

THE INSTITUTES OF JUSTINIAN, translated with
Notes by J. T. ABDY, LL.D., Judge of County Courts, late Regius
Professor of Laws in the University of Cambridge, and formerly
Fellow of Trinity Hall; and BRYAN WALKER, M.A., LL.D., Law
Lecturer of St John's College, Cambridge; late Fellow and Lecturer
of Corpus Christi College; and formerly Law Student of Trinity
Hall. Crown 8vo. 16*s.*

"We welcome here a valuable contribution to the study of jurisprudence. The text of the *Institutes* is occasionally perplexing, even to practised scholars, whose knowledge of classical models does not always avail them in dealing with the technicalities of legal phraseology. Nor can the ordinary dictionaries be expected to furnish all the help that is wanted. This translation will then be of great use. To the ordinary student, whose attention is distracted from the subject-matter by the difficulty of struggling through the language in which it is contained, it will be almost indispensable."—*Spectator.*

"The notes are learned and carefully compiled, and this edition will be found useful to students."—*Law Times.*

SELECTED TITLES FROM THE DIGEST, annotated
by B. WALKER, M.A., LL.D. Part I. Mandati vel Contra. Digest
XVII. 1. Crown 8vo. 5*s.*

"This small volume is published as an experiment. The author proposes to publish an annotated edition and translation of several books of the Digest if this one is received with favour. We are pleased to be able to say that Mr Walker deserves credit for the way in which he has performed the task undertaken. The translation, as might be expected, is scholarly."—*Law Times.*

—— Part II. De Adquirendo rerum dominio and De Adquirenda vel
amittenda possessione. Digest XLI. 1 and 11. Crown 8vo. 6*s.*

—— Part III. De Condictionibus. Digest XII. 1 and 4—7 and Digest
XIII. 1—3. Crown 8vo. 6*s.*

GROTIUS DE JURE BELLI ET PACIS, with the Notes
of Barbeyrac and others; accompanied by an abridged Translation
of the Text, by W. WHEWELL, D.D. late Master of Trinity College.
3 Vols. Demy 8vo. 12*s.* The translation separate, 6*s.*

HISTORY.

THE GROWTH OF ENGLISH INDUSTRY AND COMMERCE.

By W. CUNNINGHAM, M.A., late Deputy to the Knightbridge Professor in the University of Cambridge. With Maps and Charts. Crown 8vo. 12s.

"He is, however, undoubtedly sound in the main, and his work deserves recognition as the result of immense industry and research in a field in which the labourers have hitherto been comparatively few."—*Scotsman.*

"Mr Cunningham is not likely to disappoint any readers except such as begin by mistaking the character of his book. He does not promise, and does not give, an account of the dimensions to which English industry and commerce have grown. It is with the process of growth that he is concerned; and this process he traces with the philosophical insight which distinguishes between what is important and what is trivial. He thus follows with care, skill, and deliberation a single thread through the maze of general English history."—*Guardian.*

LIFE AND TIMES OF STEIN, OR GERMANY AND PRUSSIA IN THE NAPOLEONIC AGE,

by J. R. SEELEY, M.A., Regius Professor of Modern History in the University of Cambridge, with Portraits and Maps. 3 Vols. Demy 8vo. 48s.

"If we could conceive anything similar to a protective system in the intellectual department, we might perhaps look forward to a time when our historians would raise the cry of protection for native industry. Of the unquestionably greatest German men of modern history—I speak of Frederick the Great, Goethe and Stein—the first two found long since in Carlyle and Lewes biographers who have undoubtedly driven their German competitors out of the field. And now in the year just past Professor Seeley of Cambridge has presented us with a biography of Stein which, though it modestly declines competition with German works and disowns the presumption of teaching us Germans our own history, yet casts into the shade by its brilliant superiority all that we have ourselves hitherto written about Stein."—*Deutsche Rundschau.*

"In a notice of this kind scant justice can be done to a work like the one before us; no short *résumé* can give even the most meagre notion of the contents of these volumes, which contain no page that is superfluous, and none that is uninteresting . . . . To understand the Germany of to-day one must study the Germany of many yesterdays, and now that study has been made easy by this work, to which no one can hesitate to assign a very high place among those recent histories which have aimed at original research."—*Athenæum.*

"The book before us fills an important gap in English—nay, European—historical literature, and bridges over the history of Prussia from the time of Frederick the Great to the days of Kaiser Wilhelm. It thus gives the reader standing ground whence he may regard contemporary events in Germany in their proper historic light . . . We congratulate Cambridge and her Professor of History on the appearance of such a noteworthy production. And we may add that it is something upon which we may congratulate England that on the especial field of the Germans, history, on the history of their own country, by the use of their own literary weapons, an Englishman has produced a history of Germany in the Napoleonic age far superior to any that exists in German."—*Examiner.*

THE UNIVERSITY OF CAMBRIDGE FROM THE EARLIEST TIMES TO THE ROYAL INJUNCTIONS OF 1535,

by JAMES BASS MULLINGER, M.A. Demy 8vo. (734 pp.), 12s.

"We trust Mr Mullinger will yet continue his history and bring it down to our own day."—*Academy.*

"He has brought together a mass of instructive details respecting the rise and progress, not only of his own University, but of all the principal Universities of the Middle Ages . . . We hope some day that he may continue his labours, and give us a history of the University during the troublous times of the Reformation and the Civil War."—*Athenæum.*

"Mr Mullinger's work is one of great learning and research, which can hardly fail to become a standard book of reference on the subject . . . We can most strongly recommend this book to our readers."—*Spectator.*

VOL. II. *In the Press.*

London : Cambridge University Press Warehouse, 17 Paternoster Row.

CHRONOLOGICAL TABLES OF GREEK HISTORY.

Accompanied by a short narrative of events, with references to the sources of information and extracts from the ancient authorities, by CARL PETER. Translated from the German by G. CHAWNER, M.A., Fellow and Lecturer of King's College, Cambridge. Demy 4to. 10s.

"As a handy book of reference for genuine students, or even for learned men who want to lay their hands on an authority for some par-

ticular point as quickly as possible, the *Tables* are useful."—*Academy*.

CHRONOLOGICAL TABLES OF ROMAN HISTORY.

By the same.   [*Preparing*.

HISTORY OF THE COLLEGE OF ST JOHN THE

EVANGELIST, by THOMAS BAKER, B.D., Ejected Fellow. Edited by JOHN E. B. MAYOR, M.A., Fellow of St John's. Two Vols. Demy 8vo. 24s.

"To antiquaries the book will be a source of almost inexhaustible amusement, by historians it will be found a work of considerable service on questions respecting our social progress in past times; and the care and thoroughness with which Mr Mayor has discharged his editorial functions are creditable to his learning and industry."—*Athenæum*.

"The work displays very wide reading, and it will be of great use to members of the college and of the university, and, perhaps, of still greater use to students of English history, ecclesiastical, political, social, literary and academical, who have hitherto had to be content with 'Dyer.'"—*Academy*.

HISTORY OF NEPĀL, translated by MUNSHĪ SHEW

SHUNKER SINGH and PANDIT SHRĪ GUNĀNAND; edited with an Introductory Sketch of the Country and People by Dr D. WRIGHT, late Residency Surgeon at Kāthmāndū, and with facsimiles of native drawings, and portraits of Sir JUNG BAHĀDUR, the KING OF NEPĀL, &c. Super-royal 8vo. 21s.

"The Cambridge University Press have done well in publishing this work. Such translations are valuable not only to the historian but also to the ethnologist; ... Dr Wright's Introduction is based on personal inquiry and observation, is written intelligently and candidly, and adds much to the value of the volume. The coloured lithographic plates are

interesting."—*Nature*.

"The history has appeared at a very opportune moment...The volume...is beautifully printed, and supplied with portraits of Sir Jung Bahadoor and others, and with excellent coloured sketches illustrating Nepaulese architecture and religion."—*Examiner*.

SCHOLAE ACADEMICAE: some Account of the Studies

at the English Universities in the Eighteenth Century. By CHRISTOPHER WORDSWORTH, M.A., Fellow of Peterhouse; Author of "Social Life at the English Universities in the Eighteenth Century." Demy 8vo. 15s.

"The general object of Mr Wordsworth's book is sufficiently apparent from its title. He has collected a great quantity of minute and curious information about the working of Cambridge institutions in the last century, with an occasional comparison of the corresponding state of things at Oxford ... To a great extent it is purely a book of reference, and as such it will be of permanent value for the historical knowledge of English education and learning."—*Saturday Review*.

"Only those who have engaged in like labours will be able fully to appreciate the sustained industry and conscientious accuracy discernible in every page . . . Of the whole volume it may be said that it is a genuine service rendered to the study of University history, and that the habits of thought of any writer educated at either seat of learning in the last century will, in many cases, be far better understood after a consideration of the materials here collected."—*Academy*.

THE ARCHITECTURAL HISTORY OF THE UNI-

VERSITY AND COLLEGES OF CAMBRIDGE, by the late Professor WILLIS, M.A. With numerous Maps, Plans, and Illustrations. Continued to the present time, and edited by JOHN WILLIS CLARK, M.A., formerly Fellow of Trinity College, Cambridge.

[*In the Press*.

MISCELLANEOUS.

A CATALOGUE OF ANCIENT MARBLES IN GREAT
BRITAIN, by Prof. ADOLF MICHAELIS. Translated by C. A. M. FENNELL, M.A., late Fellow of Jesus College. Royal 8vo. Roxburgh (Morocco back), £2. 2s.

"The object of the present work of Michaelis is to describe and make known the vast treasures of ancient sculpture now accumulated in the galleries of Great Britain, the extent and value of which are scarcely appreciated, and chiefly so because there has hitherto been little accessible information about them. To the loving labours of a learned German the owners of art treasures in England are for the second time indebted for a full description of their rich possessions. Waagen gave to the private collections of pictures the advantage of his inspection and cultivated acquaintance with art, and now Michaelis performs the same office for the still less known private hoards of antique sculptures for which our country is so remarkable. The book is beautifully executed, and with its few handsome plates, and excellent indexes, does much credit to the Cambridge Press. It has not been printed in German, but appears for the first time in the English translation. All lovers of true art and of good work should be grateful to the Syndics of the University Press for the liberal facilities afforded by them towards the production of this important volume by Professor Michaelis."—*Saturday Review.*

"'Ancient Marbles' here mean relics of Greek and Roman origin which have been imported into Great Britain from classical soil. How rich this island is in respect to these remains of ancient art, every one knows, but it is equally well known that these treasures had been most inadequately described before the author of this work undertook the labour of description. Professor Michaelis has achieved so high a fame as an authority in classical archæology that it seems unnecessary to say how good a book this is."—*The Antiquary.*

LECTURES ON TEACHING, delivered in the University
of Cambridge in the Lent Term, 1880. By J. G. FITCH, M.A., Her Majesty's Inspector of Schools. Crown 8vo. New Edition. 5s.

"The lectures will be found most interesting, and deserve to be carefully studied, not only by persons directly concerned with instruction, but by parents who wish to be able to exercise an intelligent judgment in the choice of schools and teachers for their children. For ourselves, we could almost wish to be of school age again, to learn history and geography from some one who could teach them after the pattern set by Mr Fitch to his audience ... But perhaps Mr Fitch's observations on the general conditions of school-work are even more important than what he says on this or that branch of study."—*Saturday Review.*

"It comprises fifteen lectures, dealing with such subjects as organisation, discipline, examining, language, fact knowledge, science, and methods of instruction; and though the lectures make no pretention to systematic or exhaustive treatment, they yet leave very little of the ground uncovered; and they combine in an admirable way the exposition of sound principles with practical suggestions and illustrations which are evidently derived from wide and varied experience, both in teaching and in examining."—*Scotsman.*

"As principal of a training college and as a Government inspector of schools, Mr Fitch has got at his fingers' ends the working of primary education, while as assistant commissioner to the late Endowed Schools Commission he has seen something of the machinery of our higher schools ... Mr Fitch's book covers so wide a field and touches on so many burning questions that we must be content to recommend it as the best existing *vade mecum* for the teacher. ... He is always sensible, always judicious, never wanting in tact ... Mr Fitch is a scholar; he pretends to no knowledge that he does not possess; he brings to his work the ripe experience of a well-stored mind, and he possesses in a remarkable degree the art of exposition."—*Pall Mall Gazette.*

"Therefore, without reviewing the book for the second time, we are glad to avail ourselves of the opportunity of calling attention to the re-issue of the volume in the five-shilling form, bringing it within the reach of the rank and file of the profession. We cannot let the occasion pass without making special reference to the excellent section on 'punishments' in the lecture on 'Discipline.'"—*School Board Chronicle.*

THEORY AND PRACTICE OF TEACHING. By the
Rev. EDWARD THRING, M.A., Head Master of Uppingham School, late Fellow of King's College, Cambridge. Crown 8vo. 6s.

"Any attempt to summarize the contents of the volume would fail to give our readers a taste of the pleasure that its perusal has given us."—*Journal of Education.*

"In his book we have something very different from the ordinary work on education. It is full of life. It comes fresh from the busy workshop of a teacher at once practical and enthusiastic, who has evidently taken up his pen, not for the sake of writing a book, but under the compulsion of almost passionate earnestness, to give expression to his views on questions connected with the teacher's life and work. For suggestiveness and clear incisive statement of the fundamental problems which arise in dealing with the minds of children, we know of no more useful book for any teacher who is willing to throw heart and conscience, and honesty into his work."—*New York Evening Post.*

.STATUTES OF THE UNIVERSITY OF CAMBRIDGE
and for the Colleges therein, made published and approved (1878—1882) under the Universities of Oxford and Cambridge Act, 1877. With an Appendix. Demy 8vo. 16s.

THE WOODCUTTERS OF THE NETHERLANDS
during the last quarter of the Fifteenth Century. In two parts. I. History of the Woodcutters. II. Catalogue of their Woodcuts. By WILLIAM MARTIN CONWAY. [*In the Press.*

THE DIPLOMATIC CORRESPONDENCE OF EARL
GOWER, English Ambassador at the court of Versailles from June 1790 to August 1792. From the originals in the Record Office with an introduction and Notes, by OSCAR BROWNING, M.A. [*Preparing.*

A GRAMMAR OF THE IRISH LANGUAGE. By Prof.
WINDISCH. Translated by Dr NORMAN MOORE. Crown 8vo. 7s. 6d.

STATUTA ACADEMIÆ CANTABRIGIENSIS. Demy
8vo. 2s. sewed.

STATUTES OF THE UNIVERSITY OF CAMBRIDGE.
With some Acts of Parliament relating to the University. Demy 8vo. 3s. 6d.

ORDINATIONES ACADEMIÆ CANTABRIGIENSIS.
Demy 8vo. 3s. 6d.

TRUSTS, STATUTES AND DIRECTIONS affecting
(1) The Professorships of the University. (2) The Scholarships and Prizes. (3) Other Gifts and Endowments. Demy 8vo. 5s.

COMPENDIUM OF UNIVERSITY REGULATIONS,
for the use of persons in Statu Pupillari. Demy 8vo. 6d.

CATALOGUE OF THE HEBREW MANUSCRIPTS
preserved in the University Library, Cambridge. By Dr S. M. SCHILLER-SZINESSY. Volume I. containing Section I. *The Holy Scriptures;* Section II. *Commentaries on the Bible.* Demy 8vo. 9s.
Volume II. *In the Press.*

A CATALOGUE OF THE MANUSCRIPTS preserved
in the Library of the University of Cambridge. Demy 8vo. 5 Vols. 10s. each.
INDEX TO THE CATALOGUE. Demy 8vo. 10s.

A CATALOGUE OF ADVERSARIA and printed books
containing MS. notes, preserved in the Library of the University of Cambridge. 3s. 6d.

THE ILLUMINATED MANUSCRIPTS IN THE LI-
BRARY OF THE FITZWILLIAM MUSEUM, Catalogued with Descriptions, and an Introduction, by WILLIAM GEORGE SEARLE, M.A., late Fellow of Queens' College, and Vicar of Hockington, Cambridgeshire. Demy 8vo. 7s. 6d.

A CHRONOLOGICAL LIST OF THE GRACES,
Documents, and other Papers in the University Registry which concern the University Library. Demy 8vo. 2s. 6d.

CATALOGUS BIBLIOTHECÆ BURCKHARDTIANÆ.
Demy 4to. 5s.

The Cambridge Bible for Schools and Colleges.

GENERAL EDITOR: THE VERY REVEREND J. J. S. PEROWNE, D.D.,
DEAN OF PETERBOROUGH.

———◆———

THE want of an Annotated Edition of the BIBLE, in handy portions, suitable for School use, has long been felt.

In order to provide Text-books for School and Examination purposes, the CAMBRIDGE UNIVERSITY PRESS has arranged to publish the several books of the BIBLE in separate portions at a moderate price, with introductions and explanatory notes.

The Very Reverend J. J. S. PEROWNE, D.D., Dean of Peterborough, has undertaken the general editorial supervision of the work, assisted by a staff of eminent coadjutors. Some of the books have been already edited or undertaken by the following gentlemen :

- Rev. A. CARR; M.A., *Assistant Master at Wellington College.*
- Rev. T. K. CHEYNE, M.A., *Fellow of Balliol College, Oxford.*
- Rev. S. COX, *Nottingham.*
- Rev. A. B. DAVIDSON, D.D., *Professor of Hebrew, Edinburgh.*
- The Ven. F. W. FARRAR, D.D., *Archdeacon of Westminster.*
- C. D. GINSBURG, LL.D.
- Rev. A. E. HUMPHREYS, M.A., *Fellow of Trinity College, Cambridge.*
- Rev. A. F. KIRKPATRICK, M.A., *Fellow of Trinity College, Regius Professor of Hebrew.*
- Rev. J. J. LIAS, M.A., *late Professor at St David's College, Lampeter.*
- Rev. J. R. LUMBY, D.D., *Norrisian Professor of Divinity.*
- Rev. G. F. MACLEAR, D.D., *Warden of St Augustine's College, Canterbury.*
- Rev. H. C. G. MOULE, M.A., *Fellow of Trinity College, Principal of Ridley Hall, Cambridge.*
- Rev. W. F. MOULTON, D.D., *Head Master of the Leys School, Cambridge.*
- Rev. E. H. PEROWNE, D.D., *Master of Corpus Christi College, Cambridge, Examining Chaplain to the Bishop of St Asaph.*
- The Ven. T. T. PEROWNE, M.A., *Archdeacon of Norwich.*
- Rev. A. PLUMMER, M.A., D.D., *Master of University College, Durham.*
- The Very Rev. E. H. PLUMPTRE, D.D., *Dean of Wells.*
- Rev. W. SIMCOX, M.A., *Rector of Weyhill, Hants.*
- ROBERTSON SMITH, M.A., *Lord Almoner's Professor of Arabic.*
- Rev. H. D. M. SPENCE, M.A., *Hon. Canon of Gloucester Cathedral.*
- Rev. A. W. STREANE, M.A., *Fellow of Corpus Christi College, Cambridge.*

———————

London : Cambridge University Press Warehouse, 17 *Paternoster Row.*

THE CAMBRIDGE BIBLE FOR SCHOOLS & COLLEGES.

Continued.

Now Ready. Cloth, Extra Fcap. 8vo.

THE BOOK OF JOSHUA. By the Rev. G. F. MACLEAR, D.D.
With 2 Maps. 2*s.* 6*d.*

THE BOOK OF JUDGES. By the Rev. J. J. LIAS, M.A.
With Map. 3*s.* 6*d.*

THE FIRST BOOK OF SAMUEL. By the Rev. Professor
KIRKPATRICK, M.A. With Map. 3*s.* 6*d.*

THE SECOND BOOK OF SAMUEL. By the Rev. Professor
KIRKPATRICK, M.A. With 2 Maps. 3*s.* 6*d.*

THE BOOK OF ECCLESIASTES. By the Very Rev. E. H.
PLUMPTRE, D.D., Dean of Wells. 5*s.*

THE BOOK OF JEREMIAH. By the Rev. A. W. STREANE,
M.A. With Map. 4*s.* 6*d.*

THE BOOKS OF OBADIAH AND JONAH. By Archdeacon
PEROWNE. 2*s.* 6*d.*

THE BOOK OF JONAH. By Archdeacon PEROWNE. 1*s.* 6*d.*

THE BOOK OF MICAH. By the Rev. T. K. CHEYNE, M.A.
1*s.* 6*d.*

THE GOSPEL ACCORDING TO ST MATTHEW. By the
Rev. A. CARR, M.A. With 2 Maps. 2*s.* 6*d.*

THE GOSPEL ACCORDING TO ST MARK. By the Rev.
G. F. MACLEAR, D.D. With 2 Maps. 2*s.* 6*d.*

THE GOSPEL ACCORDING TO ST LUKE. By Archdeacon
F. W. FARRAR. With 4 Maps. 4*s.* 6*d.*

THE GOSPEL ACCORDING TO ST JOHN. By the Rev.
A. PLUMMER, M.A., D.D. With 4 Maps. 4*s.* 6*d.*

THE ACTS OF THE APOSTLES. By the Rev. Professor
LUMBY, D.D. With 4 Maps. 4*s.* 6*d.*

THE EPISTLE TO THE ROMANS. By the Rev. H. C. G.
MOULE, M.A. 3*s.* 6*d.*

THE FIRST EPISTLE TO THE CORINTHIANS. By the Rev.
J. J. LIAS, M.A. With a Map and Plan. 2*s.*

THE SECOND EPISTLE TO THE CORINTHIANS. By the
Rev. J. J. LIAS, M.A. 2*s.*

THE EPISTLE TO THE HEBREWS. By Archdeacon FARRAR.
3*s.* 6*d.*

THE GENERAL EPISTLE OF ST JAMES. By the Very Rev.
E. H. PLUMPTRE, D.D., Dean of Wells. 1*s.* 6*d.*

THE EPISTLES OF ST PETER AND ST JUDE. By the
same Editor. 2*s.* 6*d.*

THE EPISTLES OF ST JOHN. By the Rev. A. PLUMMER,
M.A., D.D. 3*s.* 6*d.*

London : Cambridge University Press Warehouse, 17 *Paternoster Row.*

THE CAMBRIDGE BIBLE FOR SCHOOLS & COLLEGES.
Continued.

Preparing.

THE BOOK OF GENESIS. By ROBERTSON SMITH, M.A.

THE BOOK OF EXODUS. By the Rev. C. D. GINSBURG, LL.D.

THE BOOK OF JOB. By the Rev. A. B. DAVIDSON, D.D.

THE BOOKS OF HAGGAI AND ZECHARIAH. By Archdeacon PEROWNE.

THE BOOK OF REVELATION. By the Rev. W. SIMCOX, M.A.

THE CAMBRIDGE GREEK TESTAMENT,
FOR SCHOOLS AND COLLEGES,

with a Revised Text, based on the most recent critical authorities, and English Notes, prepared under the direction of the General Editor,

THE VERY REVEREND J. J. S. PEROWNE, D.D.,

DEAN OF PETERBOROUGH.

Now Ready.

THE GOSPEL ACCORDING TO ST MATTHEW. By the Rev. A. CARR, M.A. With 4 Maps. 4s. 6d.

"With the 'Notes,' in the volume before us, we are much pleased: so far as we have searched, they are scholarly and sound. The quotations from the Classics are apt; and the references to modern Greek form a pleasing feature."—*The Churchman.*

"Copious illustrations, gathered from a great variety of sources, make his notes a very valuable aid to the student. They are indeed remarkably interesting, while all explanations on meanings, applications, and the like are distinguished by their lucidity and good sense."—*Pall Mall Gazette.*

THE GOSPEL ACCORDING TO ST MARK. By the Rev. G. F. MACLEAR, D.D. With 3 Maps. 4s. 6d.

"The Cambridge Greek Testament, of which Dr Maclear's edition of the Gospel according to St Mark is a volume, certainly supplies a want. Without pretending to compete with the leading commentaries, or to embody very much original research, it forms a most satisfactory introduction to the study of the New Testament in the original ... Dr Maclear's introduction contains all that is known of St Mark's life, with references to passages in the New Testament in which he is mentioned; an account of the circumstances in which the Gospel was composed, with an estimate of the influence of St Peter's teaching upon St Mark; an excellent sketch of the special characteristics of this Gospel; an analysis, and a chapter on the text of the New Testament generally ... The work is completed by two good maps, one of Palestine in the time of our Lord, the other, on a large scale, of the Sea of Galilee and the country immediately surrounding it."—*Saturday Review.*

"The Notes, which are admirably put together, seem to contain all that is necessary for the guidance of the student, as well as a judicious selection of passages from various sources illustrating scenery and manners."—*Academy.*

THE GOSPEL ACCORDING TO ST LUKE. By Archdeacon FARRAR. With 4 Maps. 6s.

THE GOSPEL ACCORDING TO ST JOHN. By the Rev. A. PLUMMER, M.A., D.D. With 4 Maps. 6s.

"A valuable addition has also been made to 'The Cambridge Greek Testament for Schools,' Dr Plummer's notes on 'the Gospel according to St John' are scholarly, concise, and instructive, and embody the results of much thought and wide reading."—*Expositor.*

THE PITT PRESS SERIES.

I. GREEK.

THE ANABASIS OF XENOPHON, Books I. III. IV.

and V. With a Map and English Notes by ALFRED PRETOR, M.A., Fellow of St Catharine's College, Cambridge ; Editor of *Persius* and *Cicero ad Atticum* Book I. 2*s.* each.

"In Mr Pretor's edition of the Anabasis the text of Kühner has been followed in the main, while the exhaustive and admirable notes of the great German editor have been largely utilised. These notes deal with the minutest as well as the most important difficulties in construction, and all questions of history, antiquity, and geography are briefly but very effectually elucidated."—*The Examiner.*

"We welcome this addition to the other books of the *Anabasis* so ably edited by Mr Pretor. Although originally intended for the use of candidates at the university local examinations, yet this edition will be found adapted not only to meet the wants of the junior student, but even advanced scholars will find much in this work that will repay its perusal."—*The Schoolmaster.*

"Mr Pretor's 'Anabasis of Xenophon, Book IV.' displays a union of accurate Cambridge scholarship, with experience of what is required by learners gained in examining middle-class schools. The text is large and clearly printed, and the notes explain all difficulties. . . . Mr Pretor's notes seem to be all that could be wished as regards grammar, geography, and other matters."—*The Academy.*

BOOKS II. VI. and VII. By the same Editor. 2*s.* 6*d.* each.

"Another Greek text, designed it would seem for students preparing for the local examinations, is 'Xenophon's Anabasis,' Book II., with English Notes, by Alfred Pretor, M.A. The editor has exercised his usual discrimination in utilising the text and notes of Kuhner, with the occasional assistance of the best hints of Schneider, Vollbrecht and Macmichael on critical matters, and of Mr R. W. Taylor on points of history and geography. . . When Mr Pretor commits himself to Commentator's work, he is eminently helpful. . . Had we to introduce a young Greek scholar to Xenophon, we should esteem ourselves fortunate in having Pretor's text-book as our chart and guide."—*Contemporary Review.*

THE ANABASIS OF XENOPHON, by A. PRETOR, M.A.,

Text and Notes, complete in two Volumes. 7*s.* 6*d.*

AGESILAUS OF XENOPHON. The Text revised

with Critical and Explanatory Notes, Introduction, Analysis, and Indices. By H. HAILSTONE, M.A., late Scholar of Peterhouse, Cambridge, Editor of Xenophon's Hellenics, etc. 2*s.* 6*d.*

ARISTOPHANES—RANAE. With English Notes and

Introduction by W. C. GREEN, M.A., Assistant Master at Rugby School. 3*s.* 6*d.*

ARISTOPHANES—AVES. By the same Editor. *New*

Edition. 3*s.* 6*d.*

"The notes to both plays are excellent. Much has been done in these two volumes to render the study of Aristophanes a real treat to a boy instead of a drudgery, by helping him to understand the fun and to express it in his mother tongue."—*The Examiner.*

ARISTOPHANES—PLUTUS. By the same Editor. 3*s.* 6*d.*

EURIPIDES. HERCULES FURENS. With Intro-

ductions, Notes and Analysis. By J. T. HUTCHINSON, M.A., Christ's College, and A. GRAY, M.A., Fellow of Jesus College. 2*s.*

"Messrs Hutchinson and Gray have produced a careful and useful edition."—*Saturday Review.*

THE HERACLEIDÆ OF EURIPIDES, with Introduc-

tion and Critical Notes by E. A. BECK, M.A., Fellow of Trinity Hall. 3*s.* 6*d.*

LUCIANI SOMNIUM CHARON PISCATOR ET DE
LUCTU, with English Notes by W. E. HEITLAND, M.A., Fellow of St John's College, Cambridge. New Edition, with Appendix. 3s. 6d.

OUTLINES OF THE PHILOSOPHY OF ARISTOTLE.
Edited by E. WALLACE, M.A. (See p. 30.)

II. LATIN.

M. T. CICERONIS DE AMICITIA. Edited by J. S.
REID, M.L., Fellow and Assistant Tutor of Gonville and Caius College, Cambridge. New Edition, with Additions. 3s. 6d.

"Mr Reid has decidedly attained his aim, namely, 'a thorough examination of the Latinity of the dialogue.' . . . . . The revision of the text is most valuable, and comprehends sundry acute corrections. . . . This volume, like Mr Reid's other editions, is a solid gain to the scholarship of the country."—*Athenæum*.

"A more distinct gain to scholarship is Mr Reid's able and thorough edition of the *De Amicitiā* of Cicero, a work of which, whether we regard the exhaustive introduction or the instructive and most suggestive commentary, it would be difficult to speak too highly. . . . When we come to the commentary, we are only amazed by its fulness in proportion to its bulk. Nothing is overlooked which can tend to enlarge the learner's general knowledge of Ciceronian Latin or to elucidate the text."—*Saturday Review*.

M. T. CICERONIS CATO MAJOR DE SENECTUTE.
Edited by J. S. REID, M.L. 3s. 6d.

"The notes are excellent and scholarlike, adapted for the upper forms of public schools, and likely to be useful even to more advanced students."—*Guardian*.

M. T. CICERONIS ORATIO PRO ARCHIA POETA.
Edited by J. S. REID, M.L. Revised Edition. 2s.

"It is an admirable specimen of careful editing. An Introduction tells us everything we could wish to know about Archias, about Cicero's connexion with him, about the trial, and the genuineness of the speech. The text is well and carefully printed. The notes are clear and scholar-like. . . . No boy can master this little volume without feeling that he has advanced a long step in scholarship."—*The Academy*.

M. T. CICERONIS PRO L. CORNELIO BALBO ORA-
TIO. Edited by J. S. REID, M.L. 1s. 6d.

"We are bound to recognize the pains devoted in the annotation of these two orations to the minute and thorough study of their Latinity, both in the ordinary notes and in the textual appendices."—*Saturday Review*.

M. T. CICERONIS PRO P. CORNELIO SULLA
ORATIO. Edited by J. S. REID, M.L. 3s. 6d.

"Mr Reid is so well known to scholars as a commentator on Cicero that a new work from him scarcely needs any commendation of ours. His edition of the speech *Pro Sulla* is fully equal in merit to the volumes which he has already published . . . It would be difficult to speak too highly of the notes. There could be no better way of gaining an insight into the characteristics of Cicero's style and the Latinity of his period than by making a careful study of this speech with the aid of Mr Reid's commentary . . . Mr Reid's intimate knowledge of the minutest details of scholarship enables him to detect and explain the slightest points of distinction between the usages of different authors and different periods . . . The notes are followed by a valuable appendix on the text, and another on points of orthography; an excellent index brings the work to a close."—*Saturday Review*.

M. T. CICERONIS PRO CN. PLANCIO ORATIO.
Edited by H. A. HOLDEN, LL.D., late Head Master of Ipswich School. 4s. 6d.

"As a book for students this edition can have few rivals. It is enriched by an excellent introduction and a chronological table of the principal events of the life of Cicero; while in its appendix, and in the notes on the text which are added, there is much of the greatest value. The volume is neatly got up, and is in every way commendable."—*The Scotsman*.

"Dr Holden's own edition is all that could be expected from his elegant and practised scholarship. . . . Dr Holden has evidently made up his mind as to the character of the commentary most likely to be generally useful; and he has carried out his views with admirable thoroughness."—*Academy*.

"Dr Holden has given us here an excellent edition. The commentary is even unusually full and complete; and after going through it carefully, we find little or nothing to criticize. There is an excellent introduction, lucidly explaining the circumstances under which the speech was delivered, a table of events in the life of Cicero and a useful index." *Spectator*, Oct. 29, 1881.

London : Cambridge University Press Warehouse, 17 Paternoster Row.

M. T. CICERONIS IN Q. CAECILIUM DIVINATIO

ET IN C. VERREM ACTIO PRIMA. With Introduction and Notes by W. E. HEITLAND, M.A., and HERBERT COWIE, M.A., Fellows of St John's College, Cambridge. 3*s*.

M. T. CICERONIS ORATIO PRO L. MURENA, with

English Introduction and Notes. By W. E. HEITLAND, M.A., Fellow and Classical Lecturer of St John's College, Cambridge. **Second Edition, carefully revised.** 3*s*.

"Those students are to be deemed fortunate who have to read Cicero's lively and brilliant oration for L. Murena with Mr Heitland's handy edition, which may be pronounced 'four-square' in point of equipment, and which has, not without good reason, attained the honours of a second edition."—*Saturday Review.*

M. T. CICERONIS IN GAIUM VERREM ACTIO

PRIMA. With Introduction and Notes. By H. COWIE, M.A., Fellow of St John's College, Cambridge. 1*s*. 6*d*.

M. T. CICERONIS ORATIO PRO T. A. MILONE,

with a Translation of Asconius' Introduction, Marginal Analysis and English Notes. Edited by the Rev. JOHN SMYTH PURTON, B.D., late President and Tutor of St Catharine's College. 2*s*. 6*d*.

"The editorial work is excellently done."—*The Academy.*

M. T. CICERONIS SOMNIUM SCIPIONIS. With In-

troduction and Notes. By W. D. PEARMAN, M.A., Head Master of Potsdam School, Jamaica. 2*s*.

P. OVIDII NASONIS FASTORUM LIBER VI. With

a Plan of Rome and Notes by A. SIDGWICK, M.A. Tutor of Corpus Christi College, Oxford. 1*s*. 6*d*.

" Mr Sidgwick's editing of the Sixth Book of Ovid's *Fasti* furnishes a careful and serviceable volume for average students. It eschews 'construes' which supersede the use of the dictionary, but gives full explanation of grammatical usages and historical and mythical allusions, besides illustrating peculiarities of style, true and false derivations, and the more remarkable variations of the text."—*Saturday Review.*

"It is eminently good and useful. . . . The Introduction is singularly clear on the astronomy of Ovid, which is properly shown to be ignorant and confused; there is an excellent little map of Rome, giving just the places mentioned in the text and no more ; the notes are evidently written by a practical schoolmaster."—*The Academy.*

GAI IULI CAESARIS DE BELLO GALLICO COM-

MENT. I. II. With English Notes and Map by A. G. PESKETT, M.A., Fellow of Magdalene College, Cambridge, Editor of Caesar De Bello Gallico, VII. 2*s*. 6*d*.

BOOKS III. AND VI. By the same Editor. 1*s*. 6*d*. each.

"In an unusually succinct introduction he gives all the preliminary and collateral information that is likely to be useful to a young student ; and, wherever we have examined his notes, we have found them eminently practical and satisfying. . . The book may well be recommended for careful study in school or college."—*Saturday Review.*

"The notes are scholarly, short, and a real help to the most elementary beginners in Latin prose."—*The Examiner.*

BOOKS IV. AND V. AND BOOK VII. by the same Editor.

2*s*. each.

BOOK VIII. by the same Editor.   [*In the Press.*

P. VERGILI MARONIS AENEIDOS LIBRI I., II., IV.,

V., VI., VII., VIII., IX., X., XI., XII. Edited with Notes by A. SIDGWICK, M.A. Tutor of Corpus Christi College, Oxford. 1s. 6d. each.

"Much more attention is given to the literary aspect of the poem than is usually paid to it in editions intended for the use of beginners. The introduction points out the distinction between primitive and literary epics, explains the purpose of the poem, and gives an outline of the story." —*Saturday Review.*

"Mr Arthur Sidgwick's 'Vergil, Aeneid, Book XII.' is worthy of his reputation, and is distinguished by the same acuteness and accuracy of knowledge, appreciation of a boy's difficulties and ingenuity and resource in meeting them, which we have on other occasions had reason to praise in these pages."—*The Academy.*

"As masterly in its clearly divided preface and appendices as in the sound and independent character of its annotations. . . . There is a great deal more in the notes than mere compilation and suggestion. . . . No difficulty is left unnoticed or unhandled."—*Saturday Review.*

"This edition is admirably adapted for the use of junior students, who will find in it the result of much reading in a condensed form, and clearly expressed."—*Cambridge Independent Press.*

BOOKS VII. VIII. in one volume. 3s.

BOOKS IX. X. in one volume. 3s.

BOOKS X., XI., XII. in one volume. 3s. 6d.

QUINTUS CURTIUS. A Portion of the History.

(ALEXANDER IN INDIA.) By W. E. HEITLAND, M.A., Fellow and Lecturer of St John's College, Cambridge, and T. E. RAVEN, B.A., Assistant Master in Sherborne School. 3s. 6d.

"Equally commendable as a genuine addition to the existing stock of school-books is *Alexander in India*, a compilation from the eighth and ninth books of Q. Curtius, edited for the Pitt Press by Messrs Heitland and Raven. . . . The work of Curtius has merits of its own, which, in former generations, made it a favourite with English scholars, and which still make it a popular text-book in Continental schools. . . . . . The reputation of Mr Heitland is a sufficient guarantee for the scholarship of the notes, which are ample without being excessive, and the book is well furnished with all that is needful in the nature of maps, indexes, and appendices." —*Academy.*

M. ANNAEI LUCANI PHARSALIAE LIBER

PRIMUS, edited with English Introduction and Notes by W. E. HEITLAND, M.A. and C. E. HASKINS, M.A., Fellows and Lecturers of St John's College, Cambridge. 1s. 6d.

"A careful and scholarlike production."—*Times.*

"In nice parallels of Lucan from Latin poets and from Shakspeare, Mr Haskins and Mr Heitland deserve praise."—*Saturday Review.*

BEDA'S ECCLESIASTICAL HISTORY, BOOKS

III., IV., the Text from the very ancient MS. in the Cambridge University Library, collated with six other MSS. Edited, with a life from the German of EBERT, and with Notes, &c. by J. E. B. MAYOR, M.A., Professor of Latin, and J. R. LUMBY, D.D., Norrisian Professor of Divinity. Revised edition. 7s. 6d.

"To young students of English History the illustrative notes will be of great service, while the study of the texts will be a good introduction to Mediæval Latin."—*The Nonconformist.*

"In Bede's works Englishmen can go back to *origines* of their history, unequalled for form and matter by any modern European nation. Prof. Mayor has done good service in rendering a part of Bede's greatest work accessible to those who can read Latin with ease. He has adorned this edition of the third and fourth books of the 'Ecclesiastical History' with that amazing erudition for which he is unrivalled among Englishmen and rarely equalled by Germans. And however interesting and valuable the text may be, we can certainly apply to his notes the expression, *La sauce vaut mieux que le poisson.* They are literally crammed with interesting information about early English life. For though ecclesiastical in name, Bede's history treats of all parts of the national life, since the Church had points of contact with all."—*Examiner.*

BOOKS I. and II. *In the Press.*

London : Cambridge University Press Warehouse, 17 *Paternoster Row*

III. FRENCH.

LE BOURGEOIS GENTILHOMME, Comédie-Ballet en
Cinq Actes. Par J.-B. POQUELIN DE MOLIÈRE (1670). With a life of Molière and Grammatical and Philological Notes. By the Rev. A. C. CLAPIN, M.A., St John's College, Cambridge, and Bachelier-ès-Lettres of the University of France. 1s. 6d.

LA PICCIOLA. By X. B. SAINTINE. The Text, with
Introduction, Notes and Map, by the same Editor, 2s.

LA GUERRE. By MM. ERCKMANN-CHATRIAN. With
Map, Introduction and Commentary by the same Editor. 3s.

LAZARE HOCHE—PAR ÉMILE DE BONNECHOSE.
With Three Maps, Introduction and Commentary, by C. COLBECK, M.A., late Fellow of Trinity College, Cambridge; Assistant Master at Harrow School. 2s.

HISTOIRE DU SIÈCLE DE LOUIS XIV PAR
VOLTAIRE. Part I. Chaps. I.—XIII. Edited with Notes Philological and Historical, Biographical and Geographical Indices, etc. by GUSTAVE MASSON, B.A. Univ. Gallic., Officier d'Académie, Assistant Master of Harrow School, and G. W. PROTHERO, M.A., Fellow and Tutor of King's College, Cambridge. 2s. 6d.

"Messrs Masson and Prothero have, to judge from the first part of their work, performed with much discretion and care the task of editing Voltaire's *Siècle de Louis XIV* for the 'Pitt Press Series.' Besides the usual kind of notes, the editors have in this case, influenced by Voltaire's 'summary way of treating much of the history,' given a good deal of historical information, in which they have, we think, done well. At the beginning of the book will be found excellent and succinct accounts of the constitution of the French army and Parliament at the period treated of."—*Saturday Review.*

Part II. Chaps. XIV.—XXIV. With Three Maps of the
Period. By the same Editors. 2s. 6d.

Part III. Chap. XXV. to the end. By the same Editors.
2s. 6d.

LE VERRE D'EAU. A Comedy, by SCRIBE. With a
Biographical Memoir, and Grammatical, Literary and Historical Notes. By C. COLBECK, M.A., late Fellow of Trinity College, Cambridge; Assistant Master at Harrow School. 2s.

"It may be national prejudice, but we consider this edition far superior to any of the series which hitherto have been edited exclusively by foreigners. Mr Colbeck seems better to understand the wants and difficulties of an English boy. The etymological notes especially are admirable. . . . The historical notes and introduction are a piece of thorough honest work."—*Journal of Education.*

M. DARU, par M. C. A. SAINTE-BEUVE, (Causeries du
Lundi, Vol. IX.). With Biographical Sketch of the Author, and Notes Philological and Historical. By GUSTAVE MASSON. 2s.

LA SUITE DU MENTEUR. A Comedy in Five Acts,
by P. CORNEILLE. Edited with Fontenelle's Memoir of the Author, Voltaire's Critical Remarks, and Notes Philological and Historical. By GUSTAVE MASSON. 2s.

LA JEUNE SIBÉRIENNE. LE LÉPREUX DE LA
CITÉ D'AOSTE. Tales by COUNT XAVIER DE MAISTRE. With Biographical Notice, Critical Appreciations, and Notes. By GUSTAVE MASSON. 2s.

LE DIRECTOIRE. (Considérations sur la Révolution

Française. Troisième et quatrième parties.) Par MADAME LA BARONNE DE STAËL-HOLSTEIN. With a Critical Notice of the Author, a Chronological Table, and Notes Historical and Philological, by G. MASSON, B.A., and G. W. PROTHERO, M.A. Revised and enlarged Edition. 2s.

"Prussia under Frederick the Great, and France under the Directory, bring us face to face respectively with periods of history which it is right should be known thoroughly, and which are well treated in the Pitt Press volumes. The latter in particular, an extract from the world-known work of Madame de Staël on the French Revolution, is beyond all praise for the excellence both of its style and of its matter."—*Times.*

DIX ANNÉES D'ÉXIL. LIVRE II. CHAPITRES 1—8.

Par MADAME LA BARONNE DE STAËL-HOLSTEIN. With a Biographical Sketch of the Author, a Selection of Poetical Fragments by Madame de Staël's Contemporaries, and Notes Historical and Philological. By GUSTAVE MASSON and G. W. PROTHERO, M.A. Revised and enlarged edition. 2s.

FRÉDÉGONDE ET BRUNEHAUT. A Tragedy in Five

Acts, by N. LEMERCIER. Edited with Notes, Genealogical and Chrono- · logical Tables, a Critical Introduction and a Biographical Notice. By GUSTAVE MASSON. 2s.

LE VIEUX CÉLIBATAIRE. A Comedy, by COLLIN

D'HARLEVILLE. With a Biographical Memoir, and Grammatical, Literary and Historical Notes. By the same Editor. 2s.

"M. Masson is doing good work in introducing learners to some of the less-known French play-writers. The arguments are admirably clear, and the notes are not too abundant."— *Academy.*

LA MÉTROMANIE, A Comedy, by PIRON, with a Bio-

graphical Memoir, and Grammatical, Literary and Historical Notes. By the same Editor. 2s.

LASCARIS, OU LES GRECS DU XVᴱ. SIÈCLE,

Nouvelle Historique, par A. F. VILLEMAIN, with a Biographical Sketch of the Author, a Selection of Poems on Greece, and Notes Historical and Philological. By the same Editor. 2s.

IV. GERMAN.

CULTURGESCHICHTLICHE NOVELLEN, von W. H.

RIEHL, with Grammatical, Philological, and Historical Notes, and a Complete Index, by H. J. WOLSTENHOLME, B.A. (Lond.). 4s. 6d.

ERNST, HERZOG VON SCHWABEN. UHLAND. With

Introduction and Notes. By H. J. WOLSTENHOLME, B.A. (Lond.), Lecturer in German at Newnham College, Cambridge. 3s. 6d.

ZOPF UND SCHWERT. . Lustspiel in fünf Aufzügen von

KARL GUTZKOW. With a Biographical and Historical Introduction, English Notes, and an Index. By the same Editor. 3s. 6d.

"We are glad to be able to notice a careful edition of K. Gutzkow's amusing comedy 'Zopf and Schwert' by Mr H. J. Wolstenholme. . . . These notes are abundant and contain references to standard grammatical works."—*Academy.*

Goethe's Knabenjahre. (1749—1759.) GOETHE'S BOY-

HOOD: being the First Three Books of his Autobiography. Arranged and Annotated by WILHELM WAGNER, Ph. D., late Professor at the Johanneum, Hamburg. 2s.

HAUFF. DAS WIRTHSHAUS IM SPESSART. Edited
by A. Schlottmann, Ph.D., Assistant Master at Uppingham School.
3s. 6d.

DER OBERHOF. A Tale of Westphalian Life, by Karl
Immermann. With a Life of Immermann and English Notes, by Wilhelm
Wagner, Ph.D., late Professor at the Johanneum, Hamburg. 3s.

A BOOK OF GERMAN DACTYLIC POETRY. Arranged and Annotated by the same Editor. 3s.

Der erſte Kreuzzug (THE FIRST CRUSADE), by Fried-
rich von Raumer. Condensed from the Author's 'History of the Hohen-
staufen', with a life of Raumer, two Plans and English Notes. By
the same Editor. 2s.

"Certainly no more interesting book could be made the subject of examinations. The story
of the First Crusade has an undying interest. The notes are, on the whole, good."—*Educational
Times.*

A BOOK OF BALLADS ON GERMAN HISTORY.
Arranged and Annotated by the same Editor. 2s.

"It carries the reader rapidly through some of the most important incidents connected with
the German race and name, from the invasion of Italy by the Visigoths under their King Alaric,
down to the Franco-German War and the installation of the present Emperor. The notes supply
very well the connecting links between the successive periods, and exhibit in its various phases of
growth and progress, or the reverse, the vast unwieldy mass which constitutes modern Germany."
—*Times.*

DER STAAT FRIEDRICHS DES GROSSEN. By G.
Freytag. With Notes. By the same Editor. 2s.

"Prussia under Frederick the Great, and France under the Directory, bring us face to face
respectively with periods of history which it is right should be known thoroughly, and which
are well treated in the Pitt Press volumes."—*Times.*

GOETHE'S HERMANN AND DOROTHEA. With
an Introduction and Notes. By the same Editor. Revised edition by J. W.
Cartmell, M.A. 3s. 6d.

"The notes are among the best that we know, with the reservation that they are often too
abundant."—*Academy.*

Das Jahr 1813 (THE YEAR 1813), by F. Kohlrausch.
With English Notes. By the same Editor. 2s.

V. ENGLISH.

JOHN AMOS COMENIUS, Bishop of the Moravians. His
Life and Educational Works, by S. S. Laurie, A.M., F.R.S.E., Professor of
the Institutes and History of Education in the University of Edinburgh.
Second Edition, revised. 3s. 6d.

OUTLINES OF THE PHILOSOPHY OF ARISTOTLE.
Compiled by Edwin Wallace, M.A., LL.D. (St Andrews), Fellow and
Tutor of Worcester College, Oxford. Third Edition Enlarged. 4s. 6d.

THREE LECTURES ON THE PRACTICE OF EDU-
CATION. Delivered in the University of Cambridge in the Easter Term,
1882, under the direction of the Teachers' Training Syndicate. 2s.

"Like one of Bacon's Essays, it handles those things in which the writer's life is most conver-
sant, and it will come home to men's business and bosoms. Like Bacon's Essays, too, it is full of
apophthegms."—*Journal of Education.*

GENERAL AIMS OF THE TEACHER, AND FORM
MANAGEMENT. Two Lectures delivered in the University of Cambridge
in the Lent Term, 1883, by F. W. Farrar, D.D. Archdeacon of Westminster, and R. B. Poole, B.D. Head Master of Bedford Modern School.
1s. 6d.

London : Cambridge University Press Warehouse, 17 *Paternoster Row.*

MILTON'S TRACTATE ON EDUCATION. A fac-
simile reprint from the Edition of 1673. Edited, with Introduction and Notes, by OSCAR BROWNING, M.A., Fellow and Lecturer of King's College, Cambridge, and formerly Assistant Master at Eton College. 2s.

"A separate reprint of Milton's famous letter to Master Samuel Hartlib was a desideratum, and we are grateful to Mr Browning for his elegant and scholarly edition, to which is prefixed the careful *résumé* of the work given in his 'History of Educational Theories.'"—*Journal of Education.*

LOCKE ON EDUCATION. With Introduction and Notes
by the Rev. R. H. QUICK, M.A. 3s. 6d.

"The work before us leaves nothing to be desired. It is of convenient form and reasonable price, accurately printed, and accompanied by notes which are admirable. There is no teacher too young to find this book interesting ; there is no teacher too old to find it profitable."—*The School Bulletin, New York.*

THE TWO NOBLE KINSMEN, edited with Intro-
duction and Notes by the Rev. Professor SKEAT, M.A., formerly Fellow of Christ's College, Cambridge. 3s. 6d.

"This edition of a play that is well worth study, for more reasons than one, by so careful a scholar as Mr Skeat, deserves a hearty welcome."—*Athenæum.*

"Mr Skeat is a conscientious editor, and has left no difficulty unexplained."—*Times.*

BACON'S HISTORY OF THE REIGN OF KING
HENRY VII. With Notes by the Rev. J. RAWSON LUMBY, D.D., Nor-risian Professor of Divinity ; late Fellow of St Catharine's College. 3s.

SIR THOMAS MORE'S UTOPIA. With Notes by the
Rev. J. RAWSON LUMBY, D.D., Norrisian Professor of Divinity; late Fellow of St Catharine's College, Cambridge. 3s. 6d.

"To Dr Lumby we must give praise unqualified and unstinted. He has done his work admirably. . . . Every student of history, every politician, every social reformer, every one interested in literary curiosities, every lover of English should buy and carefully read Dr Lumby's edition of the 'Utopia.' We are afraid to say more lest we should be thought extravagant, and our recommendation accordingly lose part of its force."—*The Teacher.*

"It was originally written in Latin and does not find a place on ordinary bookshelves. A very great boon has therefore been conferred on the general English reader by the managers of the *Pitt Press Series*, in the issue of a convenient little volume of *More's Utopia* not in the original Latin, but in the quaint *English Translation thereof made by Raphe Robynson*, which adds a linguistic interest to the intrinsic merit of the work. . . . All this has been edited in a most complete and scholarly fashion by Dr J. R. Lumby, the Norrisian Professor of Divinity, whose name alone is a sufficient warrant for its accuracy. It is a real addition to the modern stock of classical English literature."—*Guardian.*

MORE'S HISTORY OF KING RICHARD III. Edited
with Notes, Glossary and Index of Names. By J. RAWSON LUMBY, D.D. Norrisian Professor of Divinity, Cambridge; to which is added the conclusion of the History of King Richard III. as given in the continuation of Hardyng's Chronicle, London, 1543. 3s. 6d.

A SKETCH OF ANCIENT PHILOSOPHY FROM
THALES TO CICERO, by JOSEPH B. MAYOR, M.A., late Professor of Moral Philosophy at King's College, London. 3s. 6d.

"In writing this scholarly and attractive sketch, Professor Mayor has had chiefly in view 'undergraduates at the University or others who are commencing the study of the philosophical works of Cicero or Plato or Aristotle in the original language,' but also hopes that it 'may be found interesting and useful by educated readers generally, not merely as an introduction to the formal history of philosophy, but as supplying a key to our present ways of thinking and judging in regard to matters of the highest importance.'"—*Mind.*

"Professor Mayor contributes to the Pitt Press Series *A Sketch of Ancient Philosophy* in which he has endeavoured to give a general view of the philosophical systems illustrated by the genius of the masters of metaphysical and ethical science from Thales to Cicero. In the course of his sketch he takes occasion to give concise analyses of Plato's Republic, and of the Ethics and Politics of Aristotle ; and these abstracts will be to some readers not the least useful portions of the book. It may be objected against his design in general that ancient philosophy is too vast and too deep a subject to be dismissed in a 'sketch'—that it should be left to those who will make it a serious study. But that objection takes no account of the large class of persons who desire to know, in relation to present discussions and speculations, what famous men in the whole world thought and wrote on these topics. They have not the scholarship which would be necessary for original examination of authorities; but they have an intelligent interest in the relations between ancient and modern philosophy, and need just such information as Professor Mayor's sketch will give them."—*The Guardian.*

[Other Volumes are in preparation.]

London : Cambridge University Press Warehouse, 17 Paternoster Row.

University of Cambridge.

LOCAL EXAMINATIONS.

Examination Papers, for various years, with the *Regulations for the Examination.* Demy 8vo. 2s. each, or by Post, 2s. 2d.

Class Lists, for various years, Boys 1s., Girls 6d.

Annual Reports of the Syndicate, with Supplementary Tables showing the success and failure of the Candidates. 2s. each, by Post 2s. 3d.

HIGHER LOCAL EXAMINATIONS.

Examination Papers for 1883, *to which are added the Regulations for* 1884. Demy 8vo. 2s. each, by Post 2s. 2d.

Class Lists, for various years. 1s. By post, 1s. 2d.

Reports of the Syndicate. Demy 8vo. 1s., by Post 1s. 2d.

LOCAL LECTURES SYNDICATE.

Calendar for the years 1875—9. Fcap. 8vo. *cloth.* 2s.; for 1875—80. 2s.; for 1880—81. 1s.

TEACHERS' TRAINING SYNDICATE.

Examination Papers for various years, *to which are added the Regulations for the Examination.* Demy 8vo. 6d., by Post 7d.

CAMBRIDGE UNIVERSITY REPORTER.
Published by Authority.

Containing all the Official Notices of the University, Reports of Discussions in the Schools, and Proceedings of the Cambridge Philosophical, Antiquarian, and Philological Societies. 3d. weekly.

CAMBRIDGE UNIVERSITY EXAMINATION PAPERS.

These Papers are published in occasional numbers every Term, and in volumes for the Academical year.

VOL. X.	Parts	120 to 138.	PAPERS for the Year	1880—81,	15s. *cloth.*
VOL. XI.	„	139 to 159.	„	„	1881—82, 15s. *cloth.*
VOL. XII.	„	160 to 176.	„	„	1882—83, 15s. *cloth.*

Oxford and Cambridge Schools Examinations.

Papers set in the Examination for Certificates, July, 1882. 1s. 6d.

List of Candidates who obtained Certificates at the Examinations held in 1882 and 1883 ; and Supplementary Tables. 6d.

Regulations of the Board for 1884. 6d.

Report of the Board for the year ending Oct. 31, 1883. 1s.

London : C. J. CLAY, M.A. AND SON.
CAMBRIDGE UNIVERSITY PRESS WAREHOUSE,
17 PATERNOSTER ROW.

CAMBRIDGE : PRINTED BY C. J. CLAY, M.A. AND SON, AT THE UNIVERSITY PRESS.

www.ingramcontent.com/pod-product-compliance
Lightning Source LLC
Chambersburg PA
CBHW021945220326
41599CB00012BA/1190